TIME-RESOLVED LASER RAMAN SPECTROSCOPY

TIME-RESOLVED LASER RAMAN SPECTROSCOPY

Proceedings of a one-day symposium held by the Photochemistry Discussion Group of The Royal Institution of Great Britain, 13 December 1984

Edited by

David Phillips
The Royal Institution of Great Britain

and

George H. Atkinson
University of Arizona

harwood academic publishers
chur · london · paris · new york

Harwood Academic Publishers

Post Office Box 197
London WC2E 9PX
England

58, rue Lhomond
75005 Paris
France

Post Office Box 786
Cooper Station
New York, New York 10276
United States of America

Library of Congress Cataloging-in-Publication Data

Time-resolved laser raman spectroscopy.

Includes bibliographies and index.

1. Raman spectroscopy—Congresses. 2. Laser spectroscopy—Congresses. I. Phillips, D. (David). II. Atkinson, G. H. III. Royal Institution of Great Britain. Photochemistry Discussion Group. IV. Title: Laser raman spectroscopy.

QD96.R34T56 1986 543′.08584 86-18293

ISBN 3-7186-0343-8

 Printed in Great Britain by Bell and Bain Ltd, Glasgow.

Contents

Contributors

G. H. Atkinson	Department of Chemistry and Optical Sciences, University of Arizona, Tucson, Arizona, USA
I. Grieger	Department of Chemistry and Optical Sciences, University of Arizona, Tucson, Arizona, USA
J. P. Hardy	Department of Chemistry, Griffiths University, Brisbane, Australia
R. E. Hester	Department of Chemistry, University of York, Heslington, UK
W. J. Jones	Edward Davies Chemical Laboratories, University College of Wales, Aberystwyth, Wales, UK
P. M. Killough	Department of Chemistry, University of Arizona, Tucson, Arizona, USA
A. E. W. Knight	Department of Chemistry, Griffiths University, Brisbane, Australia
J. A. Koningstein	Department of Chemistry, Carleton University, Ottawa, Canada
J. N. Moore	Davy Faraday Research Laboratory, The Royal Institution, London, UK
D. Phillips	Davy Faraday Research Laboratory, The Royal Institution, London, UK
P. J. Rogers	Department of Chemistry, Griffiths University, Brisbane, Australia
M. Stockburger	Max-Planck Institut für biophysikalische Chemie, Göttingen, FRG

CHAPTER 1

HISTORICAL PERSPECTIVE

G.H. ATKINSON
Department of Chemistry and Optical Sciences
University of Arizona,
Tucson,
Arizona 85721

The emergence of time-resolved Raman (TR^2) spectroscopy as a major spectroscopic tool in the study of molecular structure and dynamics has been rapid. As an experimental technique which provides information on both the vibrational degrees of freedom and the kinetic behaviour of transient species, it has come to occupy an important place in molecular spectroscopy, especially for studies of the liquid phase reactions. The initial steps in the development of the instrumental methods which underlie the early TR^2 experiments were taken in the early 1970's while the first experimental studies of chemical systems were published in the mid-1970's.[1] From these beginnings, an increasingly rapid growth in the sophistication of the instrumental approaches and the depth of chemical and biochemical applications of TR^2 spectroscopy has occured. The qualitative aspects of this evolution can be seen in Figure 1 where information on the number of published papers involving TR^2 spectroscopy appearing in a given year from 1966-1984 is presented. These data are further classified by the broad categories of instrumentation (1c) and applications (1b). Although this presentation is not intended to provide a precise accounting of the published activity in this research area, it does convey the general view of rapidly increased activity, especially in the applications of TR^2 spectroscopy. Two major reasons for the expansion in applications can be found in the reliability of the instrumentation for acquiring data routinely and the steady improvement of time resolution. Progress in the latter area is evident from Figure 1a where the time resolution of TR^2 spectroscopy over the 1966-1984 period is presented.

The growth of a research field, of course, encompasses far more than the numerical expansion of research publications. The continuing development of more sophisticated measurements based on TR^2 spectroscopy is an equally important criterion which is not addressed in the information presented in Figure 1. The recent availability of:

1. time-resolved coherent Raman spectroscopies,
2. TR^2 experiments with picosecond time resolution,
3. normal coordinate analyses of large molecular systems of biochemical significance, and
4. resonance excitation in the ultraviolet spectral region are all manifestations of the continually expanding interest in TR^2 spectroscopy.

The papers presented in this volume chronical some of the more recent advances in a number of these areas. Some address instrumental advances which attempt to solve specific experimental limitations in time-resolved detection. Others address the application of well defined TR^2 techniques to significant chemical and biochemical reactions. Taken together, they represent only part of the

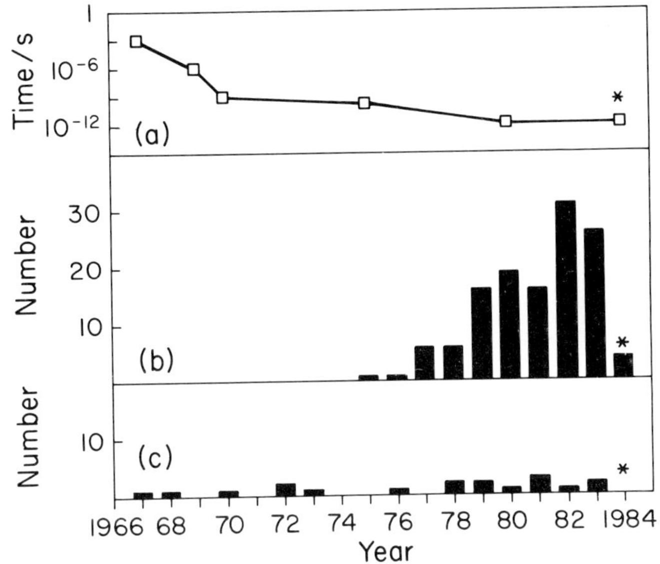

FIGURE 1 (a) The time resolution for TR^2 spectroscopy as reported in published work during the 1966-1984 period.
(b) The approximate number of papers concerning primarily the application of TR^2 spectroscopy to chemical and/or biochemical systems published in a given year during the 1966-1984 period.
(c) The approximate number of papers concerning primarily the development of new instrumental capabilities for TR^2 spectroscopy published in a given year during the 1966-1984 period. All data for 1984 pertains only to the first four months.

expanding research interest in TR^2 spectroscopy.

REFERENCES

1. see for example:

   (a) G.H. ATKINSON, Time-Resolved Raman Spectroscopy in *Advances in Infrared and Raman Spectroscopy*, Eds. R.J.N. Clark and R.E. Hester, (Heyden, London 1982), Vol. 9, pp. 1-67.

   (b) G.H. ATKINSON, Time-Resolved Raman Spectroscopy in *Advances in Laser Spectroscopy*, Eds. B.A. Garetz and J.R. Lombardi, (Heyden, London 1982), Vol. 1, pp. 155-165.

   (c) *Time-Resolved Vibrational Spectroscopy*, Ed. G.H. Atkinson, (Academic Press, New York, N.Y. 1983), pp. 1-398.

   (d) *Time-Resolved Vibrational Spectroscopy*, Ed. G.H. Atkinson, (National Science Foundation Workshop Report 1982) pp. 1-36.

CHAPTER 2

RAMAN SPECTROSCOPIC STUDIES OF TRANSIENT CHEMICAL SPECIES

R. E. HESTER
Chemistry Department,
University of York,
York YO1 5DD.

1. INTRODUCTION

There are several well-established optical techniques for monitoring chemical (including biochemical) reactions. These most commonly involve u.v./visible absorption or fluorescence in conjunction with equipment for rapid mixing of reactants (continuous or stopped-flow systems) or for inducing rapid change through photolysis, temperature-jump, pressure-jump, etc. Applied to the study of reactions in solution, such techniques can provide kinetic data for a wide variety of reaction types over time domains ranging from milliseconds for the flow methods to picoseconds for flash photolysis. Data of this type have been used widely in elucidating reaction mechanisms, including the identification of transient intermediate species. However, at best, these conventional techniques provide only a kinetic identification and some characterisation of the electronic absorption or emission spectra involved.

The recent development of vibrational Raman spectroscopic techniques coupled to fast reaction equipment has yielded important additional information. Unlike the typically broad and relatively featureless electronic spectra of solution species, vibrational spectra are highly structured and rich in detail which relates directly to molecular structure. Local modes of vibration in complex molecules commonly give rise to characteristic "group frequencies" which may be used to identify functional groups and to infer environmental influences on such groups. Delocalized, extensively coupled modes of vibration may provide complex but distinctive "fingerprints" for molecular species. For those cases where a full normal coordinate analysis is feasible, details of molecular symmetry and bond force constants may be determined from the vibrational spectra.

It has to be admitted, however, that the full potential of vibrational analysis is seldom realised for transient chemical species. In part this is due to the paucity of infrared spectroscopic data on short-lived species, but it also relates to the fact that resonance Raman spectra (RRS) are most often used to characterise transient intermediates. While resonance conditions (Raman excitation at wavelengths coincident with strong optical absorption bands) can yield spectacular enhancement in sensitivity, combined with good selectivity and thus freedom from spectral interferences, they also further limit the number of vibrational modes giving active Raman bands. Nevertheless, in spite of the difficulties, considerable progress has been made in the characterisation of structure and bonding in short lived reaction intermediates by Raman spectroscopic methods. The following few examples are chosen to illustrate this progress.

2. EXPERIMENTAL METHODS

Continuous wave (CW) laser sources are conventionally used for exciting Raman spectra. Scanning monochromators (usually double or triple systems for good stray light rejection) and high-gain, low-noise photomultipliers comprise the conventional single-channel detection systems. For use in the study of short-lived reaction intermediates such equipment must be coupled with a rapid flow system wherein the transient species of interest is continuously generated so as to establish a steady-state concentration in the sample cell.[1] Using RRS methods, minority species in reaction mixtures at solution concentrations of the order of 10^{-3} - 10^{-6} M may readily be determined.

Faster data acquisition is achieved by replacing the single-channel system specified above with a multichannel spectrometer. A spectrograph or polychromator coupled with an image-intensified vidicon or photodiode array provides an obvious multiplex advantage which has been used to good effect in combination with stopped-flow equipment for rapid mixing of solution reagents[2] and with pulse radiolysis equipment.[3] This latter technique also requires replacement of the CW laser source with a pulsed laser. With this arrangement Raman spectra are generated and detected in the time of the laser pulse duration and, since this laser pulse may be delayed with respect to the radiolysis pulse (usually an electron beam), versatile and flexible time-resolved resonance Raman spectroscopy (TR^3)

becomes possible.

The study of photochemical reaction mechanisms and identification of transient intermediates by TR^{3}s follows simply from the pulse radiolysis methods by replacing the e-beam by a second pulsed laser. Figure 1 shows such a system for flash photolysis pump-then-probe experiments. In the U.K. the Science and Engineering Research Council (SERC) recently has established a central facility of this type for university and polytechnic users at the Rutherford Laboratory (RAL) in Oxfordshire. Lambda Physik excimer and dye lasers are used in conjunction with a stimulated Raman shifter in this facility to produce line narrowed (to ca. 0.2 cm^{-1}) output at two different (tunable) wavelengths and with laser pulse widths of 5-10 ns. By means of a pulse and delay generator, used to control the triggering of lasers, two laser pulses of different wavelengths are generated simultaneously or separately with a controlled time delay for photochemical pump-then-probe experiments. For detection and analysis of the Raman scattered radiation a Spex Triplemate spectrometer fitted with an intensified multichannel diode array detector of the OMA-2 type is used. Capture of weak spectra is facilitated by repetitive pulsing of the lasers and integration of signals, with a flow system being used to change the sample between each pulse pair. For resonance Raman studies of reaction intermediates generated by methods other than photolysis only one of the laser systems is required.

3. RESULTS AND DISCUSSION

As the first example of results achieved with the RAL

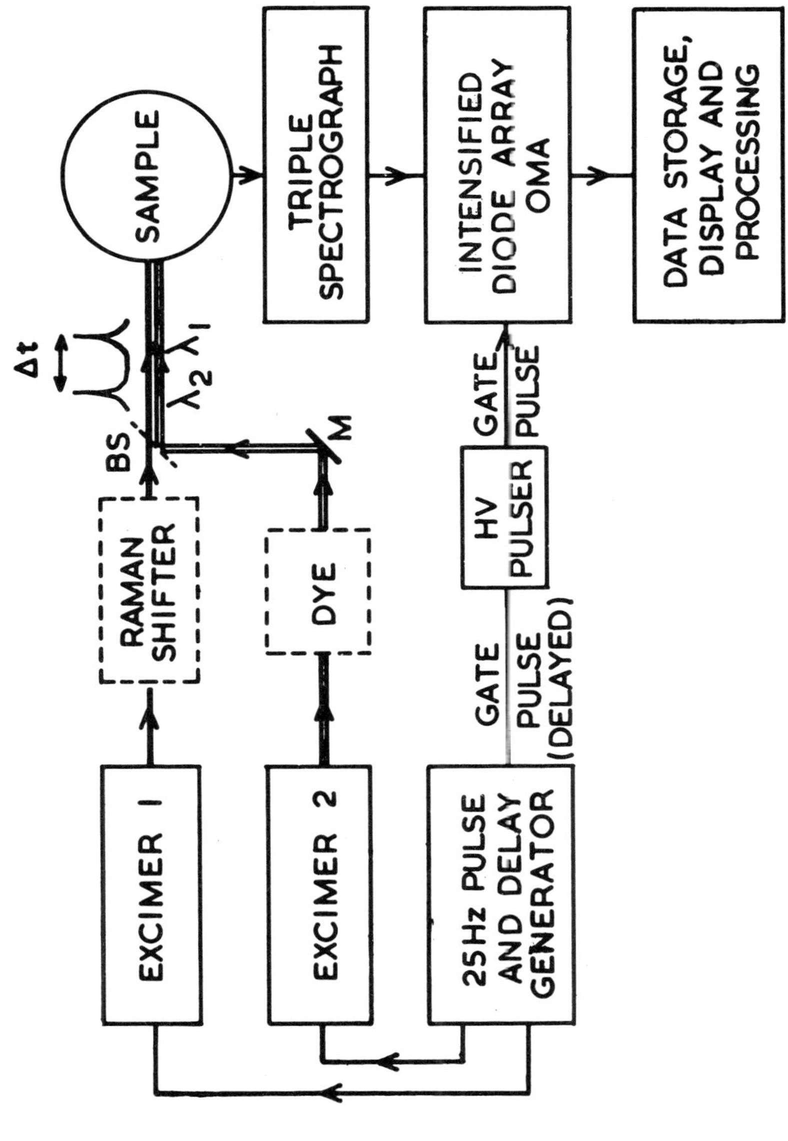

FIGURE 1 Pulsed lasers and spectrometer equipment used for Raman studies of transient species.

equipment we may consider a photocycle of anthraquinone. This system is considered in greater detail in a later article in this volume (J. N. Moore[4]), where the full reaction scheme following ultraviolet (337 nm) flash photolysis of a dilute (5×10^{-3} M) aqueous solution of anthraquinone-2,6-disulphonate (AQ26DS) is presented. The formation and subsequent decay of the free radical anion $AQ26DS^{\cdot -}$, probed via its resonance Raman spectrum, excited at 480 nm, is seen clearly in Figure 2 over the time scale from 20 ns to 10 μs after the photolysis pulse. Although the analysis of these and related data is as yet incomplete, it is immediately apparent that both kinetic[5] and structural information are yielded by these experiments. The intense resonance Raman bands at 1498 and 1347 cm^{-1} evidently are associated with CO and CC bond stretching modes which are substantially shifted to lower wavenumbers than their counterparts in the parent compound, consistent with the location of the odd electron in a π^* anti-bonding molecular orbital which is common to the two carbonyl groups and the aromatic rings.

Analysis of the spectra from the radical anion of AQ26DS is facilitated by comparison with the simpler 1,4-benzosemiquinone($SQ^{\cdot -}$) and the 4,4-biphenylsemiquinone ($BSQ^{\cdot -}$) free radical anions which also have been generated with the RAL equipment. Figure 3 shows the RR spectrum of the $BSQ^{\cdot -}$ radical anion formed from its parent hydroquinone at 10^{-2} M concentration in strongly alkaline aqueous solution. The photoreaction was pumped by the KrF 248 nm excimer laser and the RR spectrum probed with 441 nm dye laser excitation. Time delays between the two laser pulses from 20 ns to 5 μs were used and the spectrum remained strong throughout, indicating a rapid formation

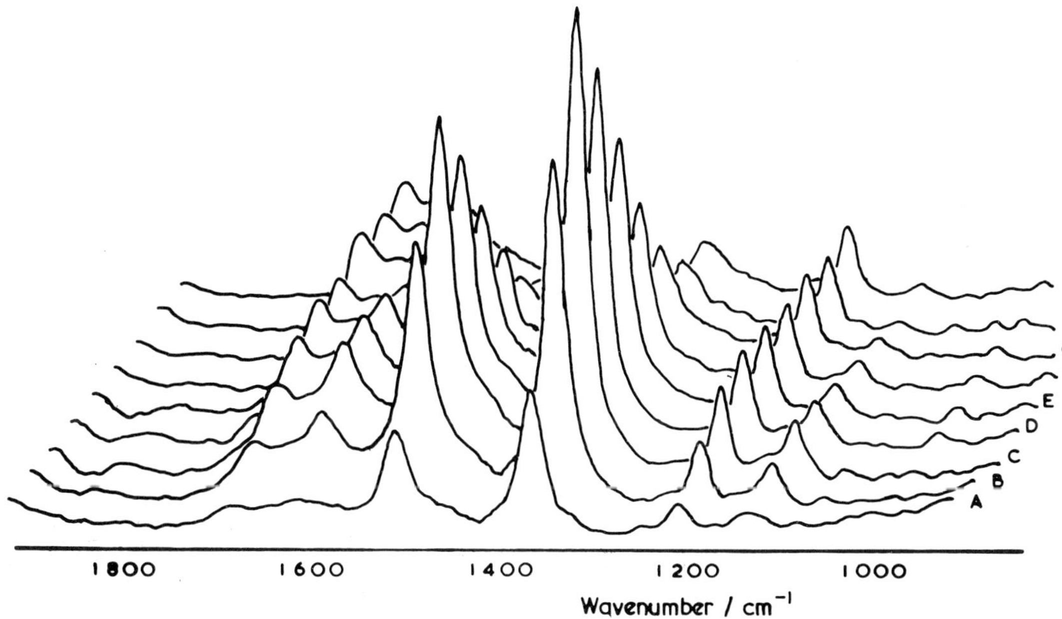

FIGURE 2 Resonance Raman spectra from 5 x 10^{-3} M anthraquinone-2, 6-disulphonate (AQ26DS) in aqueous 0.1 M sodium nitrite solution; O_2 saturated. Photolysis pump wavelength, 337 nm; Raman probe, 480 nm. Bands at 1598, 1498, 1347 and 1101 cm^{-1} are attributed to the $AQ26DS^{\cdot -}$ free radical anion.

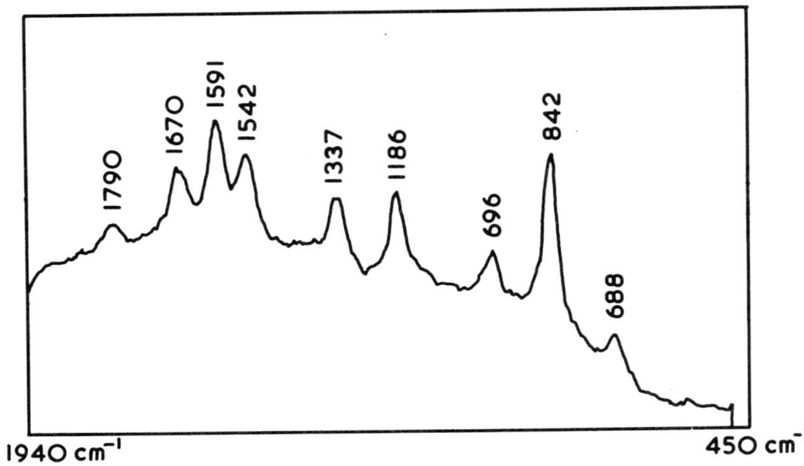

FIGURE 3 Resonance Raman spectrum of the 1,4-biphenylsemiquinone free radical anion ($BSQ^{\cdot -}$) produced by photolysis of its hydroquinone parent at 10^{-2} M in strongly alkaline aqueous solution. 248 nm pump, 20 ns delay, 441 nm Raman probe.

and good stability for the reaction intermediate. The spectrum is closely similar to that produced previously by photolysis of the hydroquinone at low temperature in glassy ethanol matrices.[6] Similar data have been obtained from benzosemiquinone[6,7] and assignments of the spectra have enabled conclusions to be drawn regarding the changes in electronic structure and chemical bonding which accompany the photoredox reactions.[6]

We have used 308 nm excimer laser radiation to generate unstable alkyl peroxy free radicals in solution, with line-narrowed and time-delayed 248 nm probe laser radiation being used to excite their resonance Raman spectra. Closely related alkyl nitroxyl radicals have served as useful model compounds in this work and spectra from two such species are shown in Figure 4. These

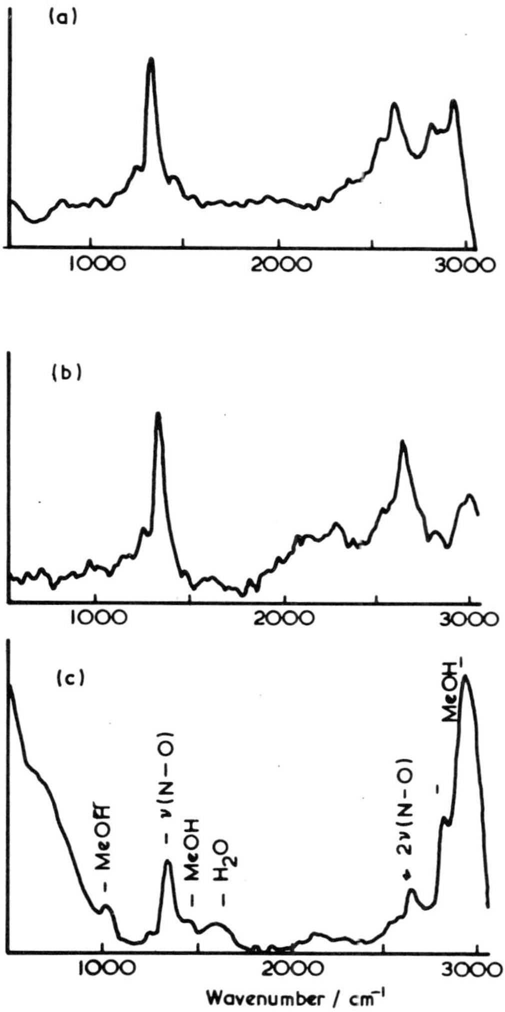

FIGURE 4 Resonance Raman spectra with excitation at 248 nm of: (a) DTBNO (5 x 10^{-3} M), solvent spectrum subtracted; (b) TEMPO (7.2 x 10^{-3} M), solvent spectrum subtracted; (c) as for (b) but without subtraction of solvent spectrum.

nitroxyl radicals (DTBNO is di-t-butyl-nitroxyl, TEMPO is 2,2,6,6-tetramethylpiperidinoxyl) are stable in the solvent used, 95% v/v water/methanol, and in spite of their structural complexity they give extremely simple RR spectra. Only one vibrational mode is resonance enhanced, this being the ν(NO) stretching mode found at ca. 1340 cm^{-1}, from which it may be deduced that the radical chromophore is highly localized in these species on the NO group. As commonly is the case for RR spectra, intense bands arising from overtones of the NO stretching mode also are observed, from which the bond vibrational anharmonicity may be calculated.[8]

A further example of our work with transient species is a recent study of the electronically excited triplet state spectrum of all-*trans*-diphenylbutadiene.[9] The shorter polyenes of this type serve as model compounds for the chromophores in larger biological systems, such as retinal in the visual pigment rhodopsin and carotenoids in caroteno-protein complexes which are widely distributed in nature and are of importance in photo-synthetic systems. Interest in these systems is focussed on their photo-isomerisation reactions, an understanding of which depends on knowledge of the nature of their photoexcited states.

Vibrational spectroscopy has been used extensively in studies of ground state structures of polyenes and, recently, through resonance Raman methods, to obtain information on excited state properties as well. From RR excitation profiles, *i.e.* the excitation wavelength dependence of RR band intensities, geometry changes between ground and excited singlet states have been determined from the spectra of butadiene,[10] hexatriene and isoprene.[11] Direct studies of the short-lived excited singlet state of stilbene also have been reported, using pump-and-probe time-resolved RR techniques.[12-14] Our own work[9] on the excited triplet state of diphenylbutadiene (DPB) involved energy transfer from a triplet sensitizer molecule, biphenyl (BP), which first was excited by 248 nm radiation. After a short time delay, the RR spectrum of the triplet DPB could be seen by probing at 396 nm.

In Figure 5 are shown transient Raman spectra obtained under various conditions of laser excitation and time delays designed to reveal the features characteristic of the excited triplet state of DPB. As is apparent from the Figure, the strong (non-resonance) spectrum of the cyclohexane solvent(s) serves as an internal intensity standard. Some pre-resonance enhancement of the DPB ground state spectrum results in the bands labelled G in the 396 nm "probe only" spectrum A. The 248 + 396 nm pump + probe spectra B-D show first of all the loss of intensity in the solvent bands due to strong absorption by the transient DPB triplet state. But in addition they show a number of new RR bands, best seen in spectrum E from which the solvent and ground state bands have been subtracted. From an analysis of these and other spectra we have been able to identify the 1582 cm^{-1} band in the transient spectrum as due to C=C double bond stretching localized in the butadiene part of the triplet state species, shifted some 43 cm^{-1} downwards from the equivalent ground state mode. The other strong band at 1556 cm^{-1} has been assigned tentatively to a ring mode shifted 39 cm^{-1} down, while the 1146 cm^{-1} band has been attributed to C-C single bond stretching localized in the butadiene part of the molecule.

Analogous experiments recently have been reported wherein 308 nm laser radiation was used to pump DPB

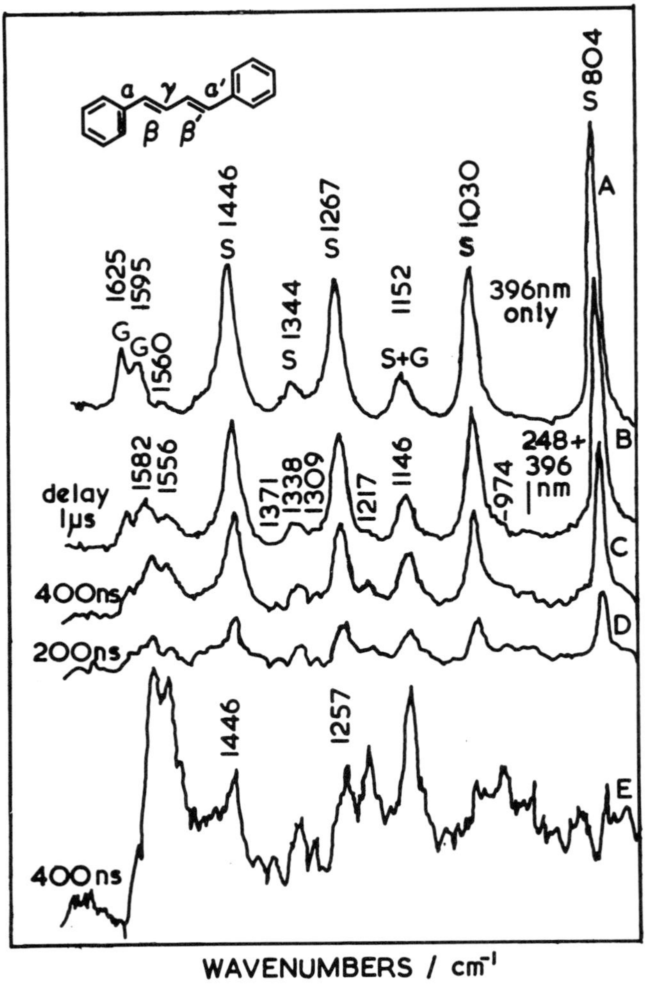

FIGURE 5 Raman spectra of a solution of 6 x 10^{-4} M DPB + 8 x 10^{-4} M BP in cyclohexane; N_2-saturated. (A) probe only, (B) - (D) pump + probe; (B) 1 s, (C) 400 ns, (D) 200 ns delay between pump and probe; (E) as for (C) after subtraction of solvent and ground state bands.

directly into its first excited singlet state, with 640 nm excitation being used to probe the RR spectrum of this species.[15] The DPB singlet state lifetime at room temperature in n-hexane is 485 ps, in cyclohexane 570 ps and in ethanol 60 ps.[16] These are much shorter than the triplet lifetime, which has been reported as 16 μs in cyclohexane at room temperature.[17] Using nanosecond lasers, therefore, the pump and probe pulses need to be coincident in time for detectable concentrations of the DPB excited singlet species to be established.[15] On the basis of $^1A_g{}^*$-symmetry in the excited singlet state, Raman bands at 1572, 1481 and 1165 cm^{-1} have been assigned to a ring mode, butadiene C=C double bond stretching and C-C single bond stretching, representing shifts of -23, -144 and +24 cm^{-1} respectively, from the ground state wavenumbers.

Diphenylbutadiene is the first molecule for which resonance Raman spectra of both the lowest excited singlet and the lowest excited triplet state, as well as the ground state, have been measured. The main spectral features are summarized in Table 1. The much larger perturbation in the butadiene C=C and C-C bonds indicated for the excited singlet state by these data is consistent with the reported observation that photoisomerization of DPB occurs mainly via a singlet state mechanism.[18]

As a final illustration of the utility of Raman spectroscopic studies for characterizing transient species we turn to photoredox reactions occurring at the surface of colloidal semiconductor hydrosols. Both TiO_2 and CdS semiconductors have been used in aqueous dispersions as photosensitizers.[19-21] Irradiation of these systems with light of wavelength short enough to promote electrons from valence bands into conduction bands results in the form-

TABLE 1 Comparison of Raman band wavenumbers (cm^{-1}) for the ground state[9] and lowest excited singlet[15] and triplet[9] states of diphenylbutadiene.

Ground	Singlet	Shift	Triplet	Shift	Assignments
1625	1481	-144	1582	-43	C=C
1595	1572	-23	1556	-39	ring
1141	1165	+24	1146	+5	C-C

ation of electron-hole (e^--h^+) pairs which may interact with surface-adsorbed species and bring about reduction and/or oxidation reactions. It has been demonstrated that adsorbed viologens are particularly effective as electron traps[19] and that phenothiazines are good positive hole traps.[19,22,23] Figure 6 shows a resonance Raman spectrum of the reduced methyl viologen species, $MV^{\cdot+}$, produced by pumping a nitrogen-purged aqueous solution of ca. 1 mM MV^{2+} in contact with a 3 mM TiO_2 sol with 248 nm KrF excimer laser radiation and, after a 100 ns delay, probing the RR spectrum with 395 nm dye laser radiation. The 248 nm radiation promotes valence electrons through the TiO_2 bandgap into the conduction band, whence they migrate rapidly to the particle surface and reduce MV^{2+}. The close similarity of the radical cation spectrum to that reported and analysed previously[24,25] for the species generated in homogeneous aqueous solution indicates that little distortion of the structure or bonding in $MV^{\cdot+}$ results from its adsorption on TiO_2. Similar conclusions may be drawn from time resolved RR studies of $MV^{\cdot+}$ on CdS sols, where the spectrum was probed only ca. 5 ns after the pump laser pulse.[26]

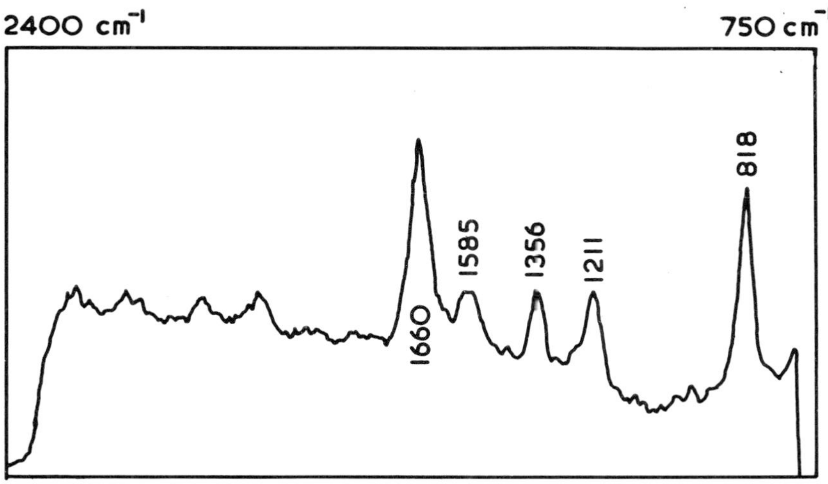

FIGURE 6 RR spectrum of the methyl viologen radical cation, $MV^{\cdot +}$, produced from ca. 1 mM MV^{2+} in 2 mM TiO_2 hydrosol at pH 11. Photo-excitation pump laser at 248 nm, RR probe laser at 395 cm^{-1}; 100 ns delay between pump and probe.

For CdS the band gap corresponds to radiation of wavelength ca. 520 nm. Thus an argon ion laser at 514.5 nm may be used in this case to simultaneously pump the semiconductor and to probe the Raman spectra of adsorbates. This wavelength yields resonance-enhanced Raman spectra for radical cations of phenothiazines[27] which typically have strong absorption bands peaking around 520 nm. Figure 7 shows Raman spectra excited at 514.5 nm from [4-(phenothiazin-10-yl)butyl]trimethylammonium chloride, 10-BPTZ, in the presence (A) and absence (B) of a CdS hydrosol. The striking difference between these two spectra has been

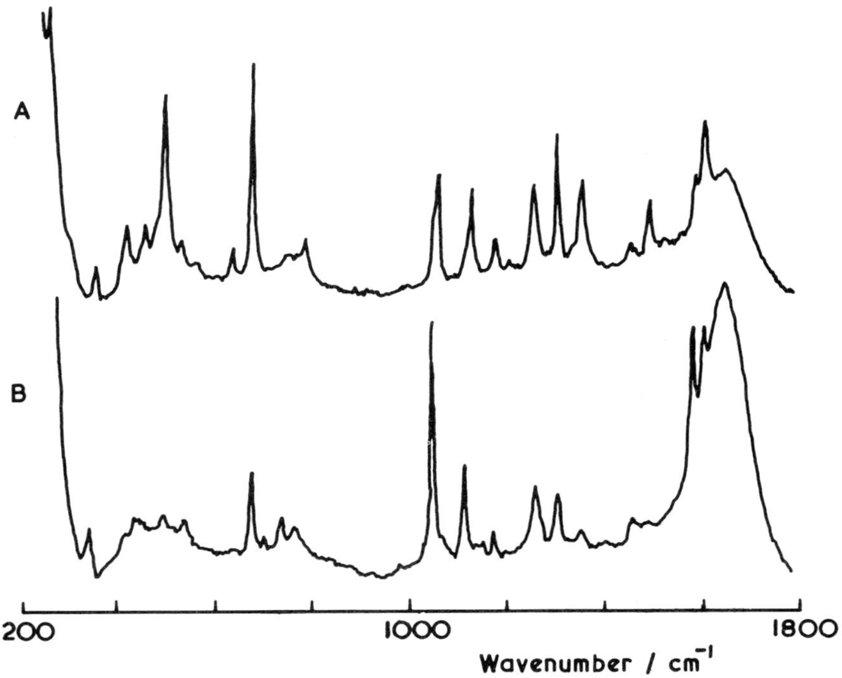

FIGURE 7 Raman spectra of 10-BPTZ. (A) ca. 2 mM 10-BPTZ in 5 mM CdS hydrosol, stabilized by 5 mM CTAB: (B) 0.1 M aqueous 10-BPTZ. 514.5 nm excitation.

attributed to efficient positive hole scavenging by the CdS surface-adsorbed phenothiazine derivative, resulting in rapid formation of the $(10\text{-BPTZ})^{\cdot +}$ radical cation.[23] Detailed analysis of these complex spectra has been made[23] by comparison with related compounds.[27]

REFERENCES

1. R. E. HESTER, in Advances in Infrared and Raman Spectroscopy, Vol. 4, Eds. R. J. H. Clark and R. E. Hester, Wiley, Chichester, (1978); E. E. ERNSTBRUNNER, R. B. GIRLING, W. E. L. GROSSMAN, E. MAYER, K. P. J. WILLIAMS and R. E. HESTER, J. Raman Spectrosc. 10, 161, (1981); R. E. HESTER and K. P. J. WILLIAMS, J. Chem. Soc. Faraday Trans., 2, 78, 573, (1982).
2. J. C. MERLIN, J. L. LORRIAUX and R. E. HESTER, J. Raman Spectrosc., 11, 384, (1981); H. SHINDO and J. HIRAISHI, Chem. Phys. Lett., 80, 238, (1981).
3. R. WILBRANDT, P. PAGSBERG, K. B. HANSEN and K. V. WEISBERG, Chem. Phys. Lett., 36, 76, (1975); R. WILBRANDT, N. H. JENSEN, P. PAGSBERG, A. SILLESEN, K. B. HANSEN and R. E. HESTER, Chem. Phys. Lett., 60, 315, (1979); K. B. HANSEN, R. WILBRANDT and P. PAGSBERT, Rev. Sci. Instrum., 50, 50, (1979).
4. J. N. MOORE, this volume ch.
5. J. N. MOORE, G. H. ATKINSON, D. PHILLIPS, P. M. KILLOUGH and R. E. HESTER, Chem. Phys. Lett., 107, 381 (1984).
6. R. E. HESTER and K. P. J. WILLIAMS, J. Chem. Soc. Faraday Trans., 2, 78, 573, (1982).
7. S. M. BECK and L. E. BRUS, J. Am. Chem. Soc., 104, 1103, 4789, (1982).
8. J. E. BENNETT, W. E. L. GROSSMAN, R. E. HESTER and P. M. KILLOUGH, J. Raman Spectrosc., in press (1985).
9. R. WILLBRANDT, W. E. L. GROSSMAN, P. M. KILLOUGH, J. E. BENNETT and R. E. HESTER, J. Phys. Chem., 88, 5964, (1984).
10. R. J. HEMLEY, J. I. DAWSON and V. VOIDA, J. Chem. Phys., 78, 2915, (1983).
11. A. B. MYERS, R. A. MATHIES, D. J. TANNOR and E. J. HELLER, J. Chem. Phys., 77, 3857, (1982).
12. T. L. GUSTAFGSON, D. M. ROBERTS and D. A. CHERNOFF, J. Chem. Phys., 79, 1559, (1983).
13. H. HAMAGUCHI, C. KATO and M. TASUMI, Chem. Phys. Lett. 100, 3, (1983).
14. H. HAMAGUCHI, T. URANO and M. TASUMI, Chem. Phys. Lett., 106, 153, (1984).
15. R. WILBRANDT, N. H. JENSEN and F. W. LANGKILDE, Chem. Phys. Lett., 111, 123, (1984).
16. S. P. VELSO and G. R. FLEMING, J. Chem. Phys., 76, 3553, (1982).
17. S. K. CHATTOPADHYAY, P. K. DAS and G. L. HUG, J. Am. Chem. Soc., 104, 4507, (1982).

18. H. GORNER, J. Photochem. 19, 343, (1982).
19. D. DUONGHONG, J. RAMSDEN and M. GRATZEL, J. Am. Chem. Soc., 104, 2977, (1982).
20. R. ROSETTI, S. M. BECK and L. E. BRUS, J. Am. Chem. Soc., 104, (1982).
21. K. METCALFE and R. E. HESTER, J. Chem. Soc. Chem. Commun., 133, (1983).
22. J. J. RAMSDEN, Thesis, EPFL Lausanne, (1985).
23. W. E. L. GROSSMAN and R. E. HESTER, J. Phys. Chem., (1985).
24. M. FORSTER, R. B. GIRLING and R. E. HESTER, J. Raman Spectrosc., 12, 36, (1982).
25. R. E. HESTER and S. SUZUKI, J. Phys. Chem., 86, 4626, (1982).
26. R. ROSETTI, S. M. BECK and L. E. BRUS, J. Am. Chem. Soc., 106, 980, (1984).
27. R. E. HESTER and K. P. J. WILLIAMS, J. Chem. Soc. Perkin Trans., 2, 852, (1981).

CHAPTER 3

TIME-RESOLVED RESONANCE RAMAN SPECTROSCOPY OF ELECTRONICALLY EXCITED SPECIES

J.N. MOORE and D. PHILLIPS,
The Royal Institution,
21 Albemarle Street,
London. W1X 4BS

In earlier papers in this volume the applications of time-resolved resonantly enhanced Raman spectroscopy to the study of a variety of molecular species has been discussed. In this brief review, we concentrate upon the time-dependent vibrational spectroscopy of electronically excited states. The type of data to be obtained can be illustrated with the example of chrysene, in which the ^3B_u state, populated by intersystem crossing from the $^1B_{3u}$ level initially populated by 354.7 nm radiation (Figure 1) yields a time-dependent vibrational spectrum, (Figure 2).[1,2]

Reviews of time-dependent Raman spectra obtained up until 1982 have appeared,[1,3] and we concentrate here upon more recent papers.

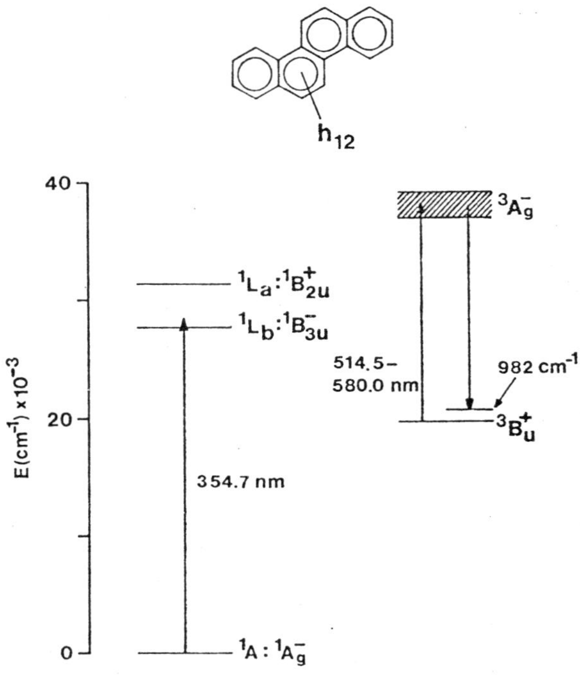

FIGURE 1 Electronic energy levels in chrysene (Atkinson and Dosser[2]).

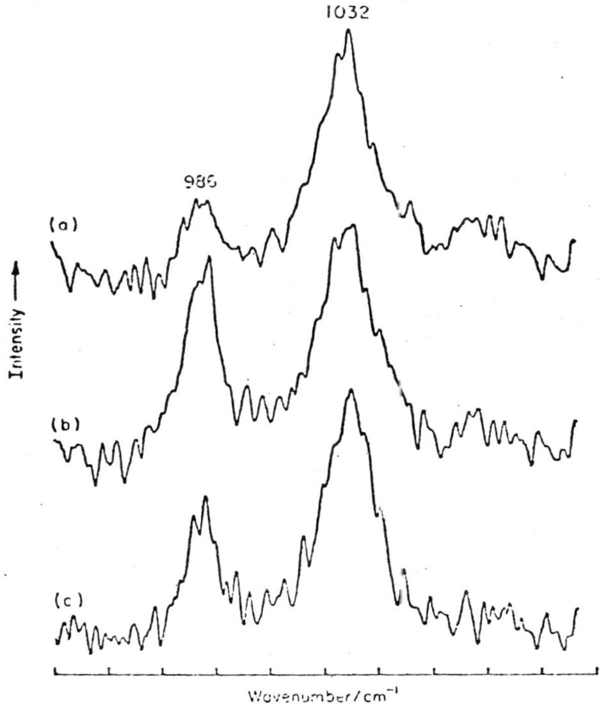

FIGURE 2 Transient resonance Raman spectrum of ^3B_u chrysene obtained at time delays between the pump and probe lasers of (a) 1500 ns, (b) 200 ns and (c) 35 ns. The band at 982 cm^{-1} is assigned to ^3B_u chrysene. The conditions used to obtain these spectra were 10^{-3} M chrysene in THF at room temperature with 354.7 nm as the pump laser wavelength and 570 nm the probe laser wavelength. (Atkinson[1]).

1. POLYENES

1.1. Stilbene

Several recent papers have addressed the problem of the structure of electronically excited states involved in the photoisomerization of stilbene.[4-7] As an example of a nanosecond study, the transient spectrum of trans-stilbene in n-hexane is shown in Figure 3,[4,5] and that for deuterated analogues in Figure 4.[5] Analysis of these spectra led to assignments for the S_1 and S_o vibrations shown in Table 1.

TABLE I Frequencies (cm^{-1}) and assignments of vibrations in S_o, S_1 states of trans-stilbene.[4,5]

S_1			S_o			Assignment
tSB-d_o	-d_2	-d_{10}	tSB-d_o	-d_2	-d_{10}	
1566	1527	1562	1640	1600	1640	C=C stretch
1241	982	1229	1318	1016	1314	olefinic CH in-plane bend
1148	1147	1147	1193	1191	1148	C-phenyl stretch

A more complete study using picosecond time-resolution has appeared recently, an example of this work being given in Figure 5.[6,7]

This leads to the assignments shown in Table 2. On the timescale of the experiments, the shape of the spectrum does not change, suggesting that vibrational relaxation is complete very quickly, i.e. within 25 ps. The authors recognise that the assignments are approximate, and that a detailed normal analysis is

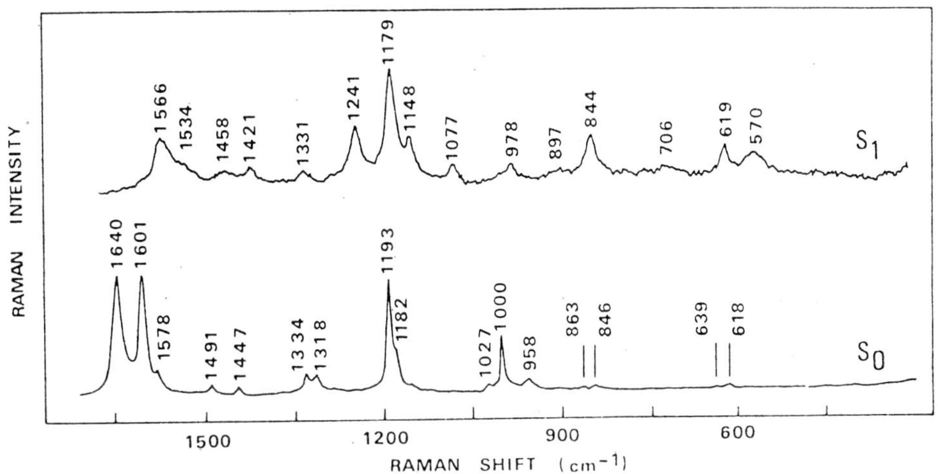

FIGURE 3 Transient resonance Raman spectrum of the S_1 state and Raman spectrum of the S_0 state of trans-stilbene obtained using pump and probe laser wavelengths of 266 nm and 585 nm respectively, pulsewidths 5n. (Hamaguchi et al.[5]).

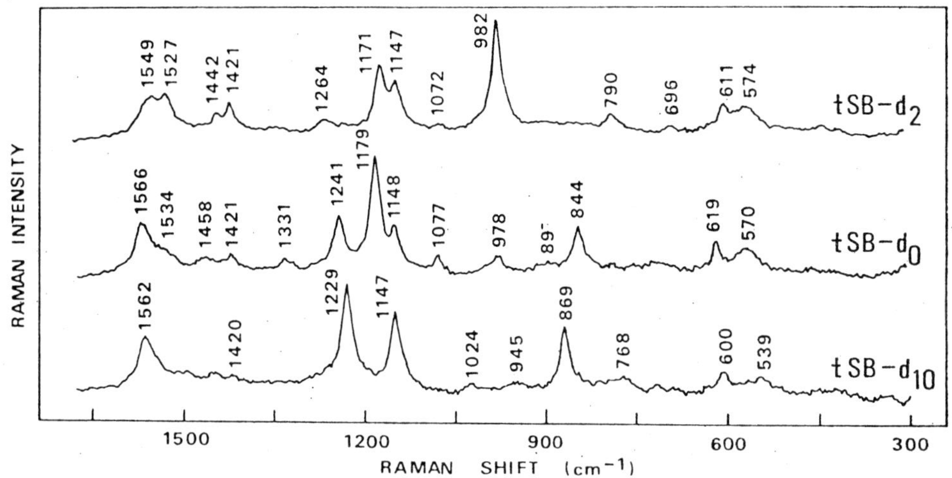

FIGURE 4 Transient resonance Raman spectra of the S_1 states of trans-stilbene ($tSB\text{-}d_o$) and two deuterated analogues ($tSB\text{-}d_2$) and ($tSB\text{-}d_{10}$). (Hamaguchi et al.[5]).

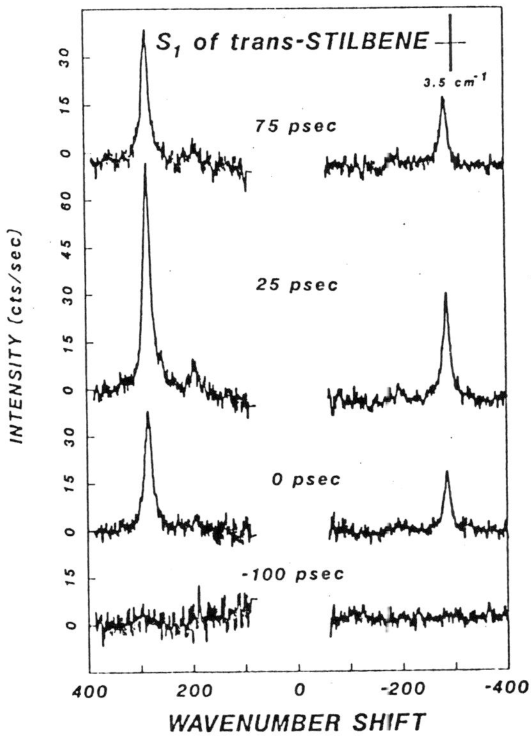

FIGURE 5

Stokes and anti-Stokes transient resonance Raman spectra of trans-stilbene obtained using a probe wavelength of 592.7 nm at various time delays following excitation at 296.3 nm, pulsewidth 25 ps. (Gustafson et al)[7]

TABLE II Frequencies (cm^{-1}) of observed modes in S_1 trans-stilbene.[7]

$\bar{\nu}$	Symmetry	Approximate assignment	Band shape
197	a_g	C_o-C_o-C bend	sharp
284	a_g	C_o-C-C bend	sharp
456	a_u(?)	ring def.	broad
565	a_g	C-C-C bend	broad
621	a	ring def.	sharp
723	a_u(?)	ring def.	broad
790	a_u(?)	C-H def.	broad
845	a_g	ring breath	sharp
979	a_g	ring def. (trigonal)	sharp
1078	a_g	C-C-H bend	sharp
1150	a_g	C-C-H bend	sharp
1181	a_g	C-H def.	sharp
1242	a_g	C_o-Ø stretch	broad, asym.
1333	a_g	C_o-C_o-H bend	sharp
1422	a	C-C stretch + C-C-H bend	sharp
1473	a_g	C-C stretch	broad
1567	a_g	C_o-C_o stretch	broad, asym.
1611	a_g	C-C stretch	broad

required before geometry can be commented upon.

1.2. Diphenylbutadiene

The vibrational spectrum of the triplet state of all-trans diphenylbutadiene has been recorded by nanosecond pump and probe methods.[8] The results show clearly that the double-bond stretching mode of the butadiene is shifted by 43 cm^{-1} down from that in the ground state, whereas the single bond stretching mode frequency is essentially unchanged. A transient band at 1556 cm^{-1} was tentatively assigned to the 8a ring mode, shifted downwards by some 39 cm^{-1}. Results are summarized in Table 3.

TABLE III Frequencies (cm^{-1}) and assignments of Diphenylbutadiene in S_1, T_1 states.[8]

Singlet	Shift	Triplet	Shift	Assignments/ comments
1620	-5	1582 s	-43	C_b=C_b str
1607	+12	1556 s	-39	8a, ring str
1492				
		1371 vw (?)		doubtful
1330		1338 w		
		1309 vw (?)		doubtful
		1256 (?)		possible artifact, doubtful
1240				
		1217 m		
1177				
1146	+5	1146 s	+5	C_b-C_b str, C_b-C str
981	-19	974 vw	-26	12, ring def.

1.3. Retinals

Time-resolved resonance Raman spectra have been obtained for the triplet states of all-trans-retinal,[9-13] 9-cis-retinal,[9,10,13] the structure being shown in Scheme 1.

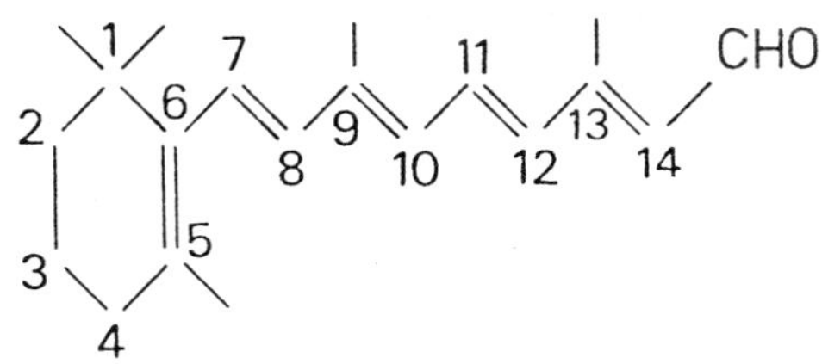

Scheme 1

In the latest study,[13] the transient Raman spectra of the triplet state produced by sensitization using the triplet state of anthracene or naphthalene as a donor yielded the results shown in Table 4, which are in essential agreement with those produced directly.[10] It is clear from Figure 6 and Table 4 that the energy of the sensitizer does not alter the Raman spectrum. Moreover, with the exception of the 13-cis isomer, this results in a common spectrum. This suggests[10] that the photoisomerization in this molecule proceeds as follows. Photoexcitation of any mono-cis isomer except 13-cis results, after some relaxation processes, in 'cis'-T_1 which subsequently relaxes either to cis-S_0 or to all-trans T_1, which is the most stable isomer of T_1 retinal. The fact that the 13-cis isomer behaves differently is probably due to the unique character of the C_{13}-C_{14} double bond, being the one which undergoes the trans-to-cis isomerization.

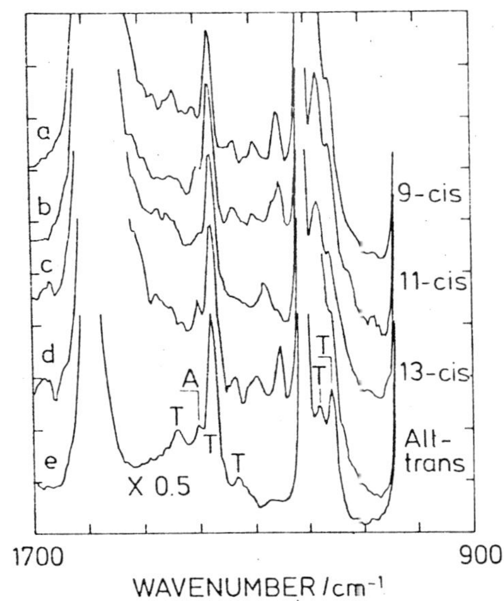

FIGURE 6 Transient resonance Raman spectra of isomers of retinal obtained following electron irradiation in the presence of anthracene as sensitizer. Bands from toluene denoted T, from anthracene denoted A. (a) 9-cis-, (b) 11-cis-, (c) 13-cis, (d) all-trans-retinal, and (e) all-trans-retinal without electron irradiation. Probe laser wavelength 480.0 nm. (Wilbrandt et al[13]).

TABLE IV Frequencies (cm^{-1}) of transient Raman bands of isomers of retinal in their triplet states sensitized with anthracene (A) or naphthalene (N).[13]

9-cis	11-cis		13-cis		all-trans	
A	A	N	A	N	A	N
1081	1080	1084	1090	1081		
1184	1183	1182	1182	1181	1184	1184
1254	1254	1255			1254	1254
			1280	1282		
1298	1304	1312			1296	1304
1334	1341	1339			1338	1340
1554	1555	1555	1550	1551	1552	1553

1.4 Carotenoids

Earlier work investigated the vibrational spectra of the long-lived triplet state of β-carotene,[14,15] zeaxanthin, echinenone, canthaxanthin, dihydroxylcopene, astaxanthin, decapreno(C_{50})-β-carotene, β-apo-8'-carotenal, and ethyl β-apo-8'-carotenoate[16] (Scheme 2). A typical early result is shown in Figure 7 for canthaxanthin,[16] and a summary of the Raman frequencies obtained for the compounds in Scheme 2 are summarized in Table 5.[16] Koningstein has recently investigated β-carotene further,[17-19] obtaining excited singlet state spectra.

For all-trans-β-carotene in carbon disulphide solution, the electronic origin at 19,480 cm^{-1} was assigned on the basis of laser Raman techniques to the 0-0 transition in the $S_1 \leftarrow S_0$ system. Another origin at 19, 794 cm^{-1} was tentatively assigned either to the 0-0 transition of an unidentified photoisomer, or to a second

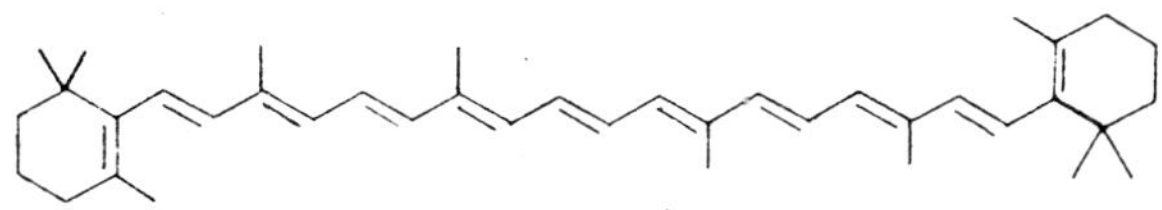

β-CAROTENE

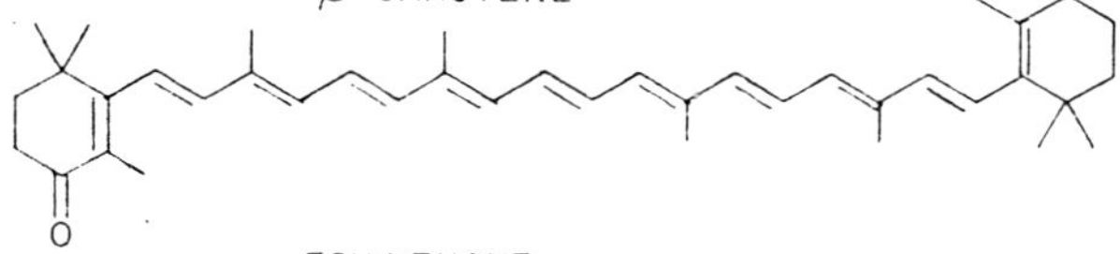

ECHINENONE

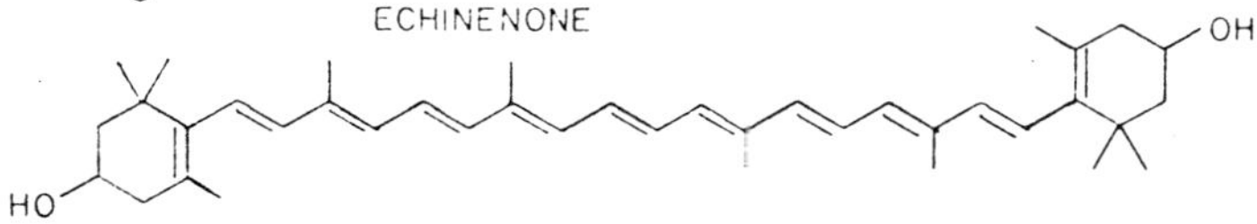

ZEAXANTHIN

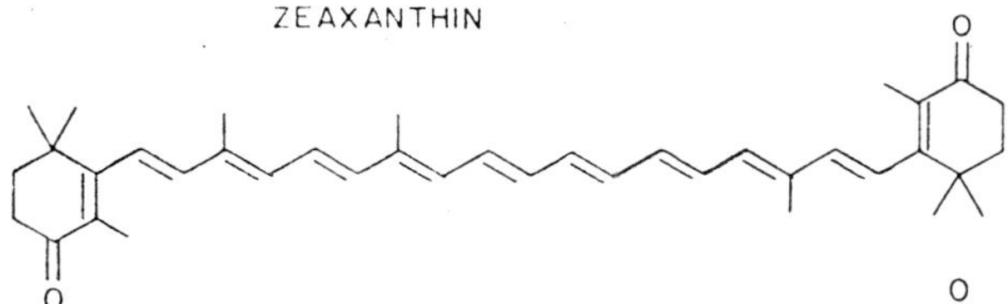

CANTHAXANTHIN

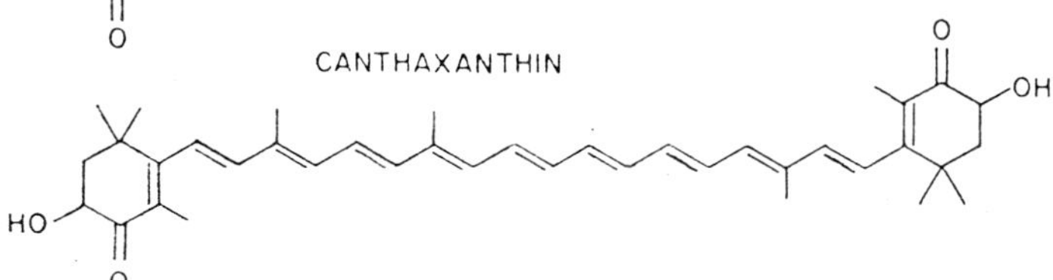

ASTAXANTHIN

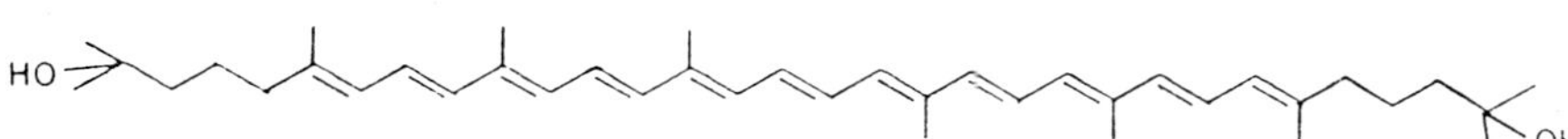

DIHYDROXYLYCOPENE

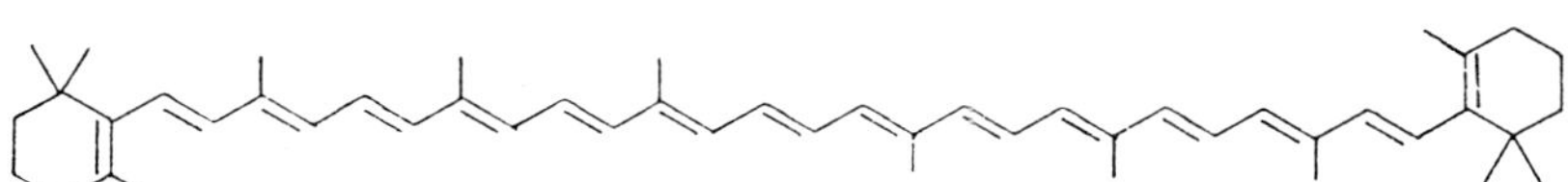

DECAPRENO-β-CAROTENE

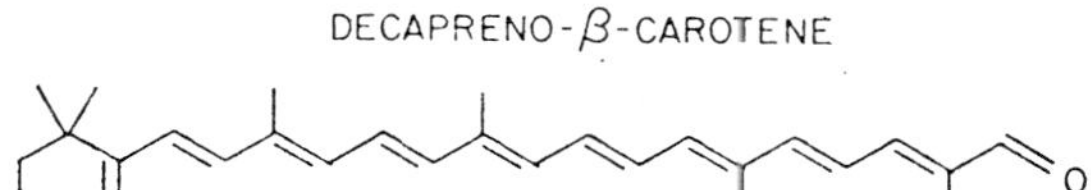

β-apo-8'-CAROTENAL

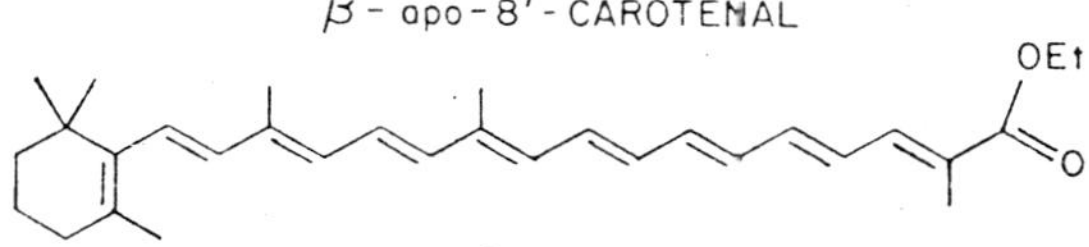

ETHYL-β-apo-8'-CAROTENOATE

Scheme 2

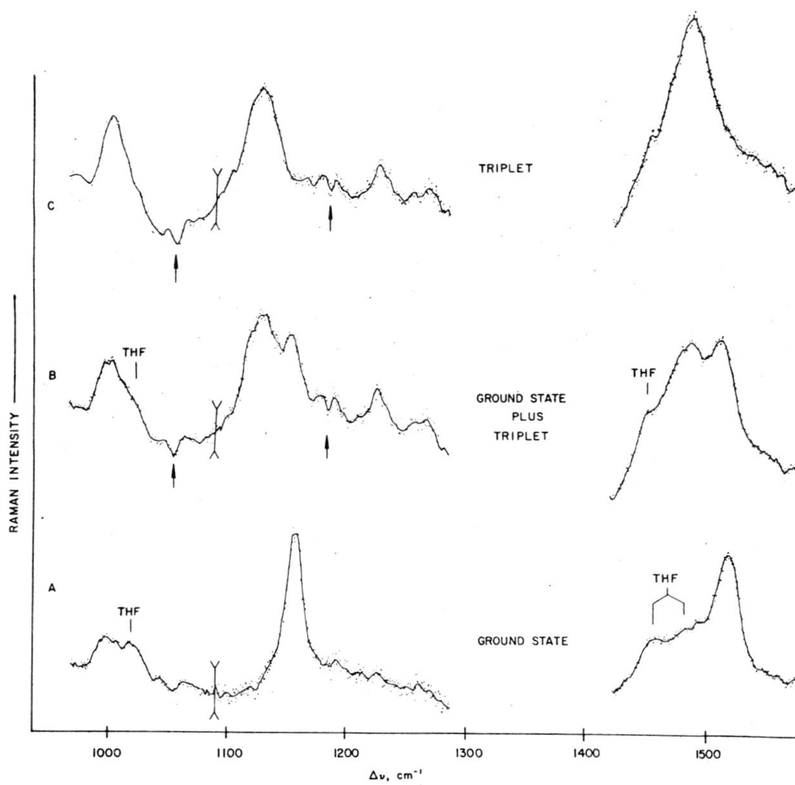

FIGURE 7
Resonance Raman spectra of canthaxanthin in the S_o and T_1 states obtained using a probe laser wavelength of 555 nm. (a) Probe laser alone, (b) transient spectrum obtained 2 μs after excitation of anthracene sensitizer, (c) spectrum obtained by normalised subtraction of (a) and (b) and assigned to T_1. Vertical arrows indicate OMA artifacts. (Dallinger et al.[16]).

TABLE V Vibrational frequences (cm^{-1}) of ground and T_1 states of carotenoids.[16]

carotenoid	Band A S_0 T_1	Band A Δ	Band B S_0 T_1	Band B Δ	Band C S_0 T_1	Band C Δ	Band D S_0 T_1	Band D Δ	Band E S_0 T_1	Band E Δ	Band F S_0 T_1	Band F Δ
β-carotene	1003 1014	+11	1157 1126	-31	1190 1188	-2	1211 1236	+25	1266		1521 1498	-23
echinenone	1004 1011	+7	1156 1127	-29	1192		1213		1272		1521 1493	-28
zeaxanthin	 1014		1156 1123	-33	1190		1212		1266		1522 1501	-21
canthaxanthin	1003 1011	+8	1155 1129(1119)	-26	1192		1214 1225	+11	1270		1519 1491(1484)	-28
astaxanthin	1005 1016	+11	1155 1128	-29	1192		1215		1277		1519 1493	-26
dihydroxylycopene	1001 1007	+6	1153 1127	-26	1196		1214		1280		1511 1490	-21
decapreno-β-carotene			1151 1124	-26	1194		1212		1283		1504 1484	-27
β-Apo-8'-carotenal	 1014		1158 1134	-24	1183		1213		1269		1527 1512	-15
ethyl β-apo-8'-carotenoate	 1012		1158 1138	-20	1193 1185	-8	1209		1274 1257	-17	1528 1510	-18

singlet-singlet transition of all-trans-β-carotene.[17] The S_1 state lifetime of β-carotene was assessed through broadening of bands to be 265 fs. Vibrational frequencies and bandwidths are shown in Table 6.

TABLE VI Vibrational frequencies and bandwidths of ground state and S_1 state of β-carotene.[18]

S_o		S_1		
$\Delta\bar{\nu}(cm^{-1})$	$\Delta\bar{\nu}(cm^{-1})$	$\Delta\bar{\nu}(cm^{-1})$	$\Delta\bar{\nu}(cm^{-1})$	
1524	21 ± 1	1524	34 ± 1	C-C in phase
1214	-	1214	28 ± 4	
1193	20 ± 2	1193	28 ± 3	
1157	19 ± 1	1159	28 ± 1	(C-C in phase)
1130	19 ± 2	1130	26 ± 5	
1003	29 ± 3	1003	40 ± 2	C-H in plane

It is interesting that with the exception of the 1157 cm^{-1} band, the frequencies are identical in S_o and S_1. A comparison of time-resolved Raman spectra of 15,15'-cis-β-carotene and 15,15'-trans-β-carotene in CS_2 and hexane reveals[19] that photoisomerization occurs in CS_2, but not in hexane. These results are discussed further in another article in this volume.

1.5 Bacteriorhodopsin

There are several papers in the present volume which discuss the transient vibrational spectra of the bacteriorhodopsin systems, and thus we merely cite here earlier work from pioneers in this field. Rhodopsin, the photoreceptor pigment in vertebrate retinal rod cells and consisting of 11-cis retinal bound to opsin, a 38-kilodalton protein was shown by picosecond resonance

Raman techniques to produce vibrational spectra typical of a distorted all-trans retinal within 30 ps of pulsed irradiation.[20] Similar fast processes were shown to occur in bacteriorhodopsin.[21,22,23]

2. OTHER PHOTOISOMERIZATIONS

In the discussion of polyenes above, vibrational spectroscopy has been shown to be useful in the study of geometrical isomerizations. One further similar study has appeared, in which the trans to cis photoisomerization of N-methylthioacetamide (Scheme 3) was studied.[24] The limit limiting timescale of these experiments was 5 ns, and it

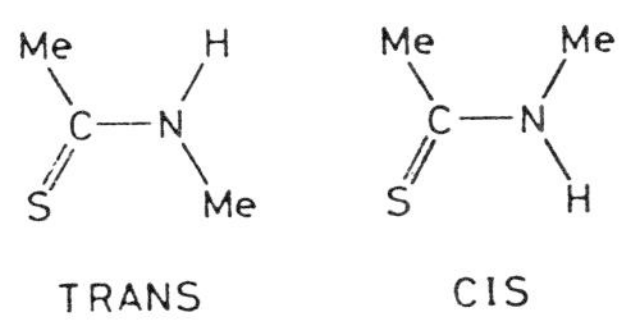

Scheme 3

was shown that the isomerization was complete within this timescale. The most likely route to isomerization is via intersystem crossing from the $^1\pi\pi^*$ state to the $^3n\pi^*$. although the possibility of direct isomerization from the $^1\pi\pi^*$ could not be eliminated entirely.

3. REACTIVE EXCITED STATES

Transient resonance Raman methods have been used to good effect in the study of the reaction dynamics of several reactive excited states. In the first example, a sequence or previously unobserved protonation reactions were observed in aqueous excited triplet state quinoxaline (1,4-diazanaphthalene).[25] These reactions were not observed in earlier transient absorption experiments. A

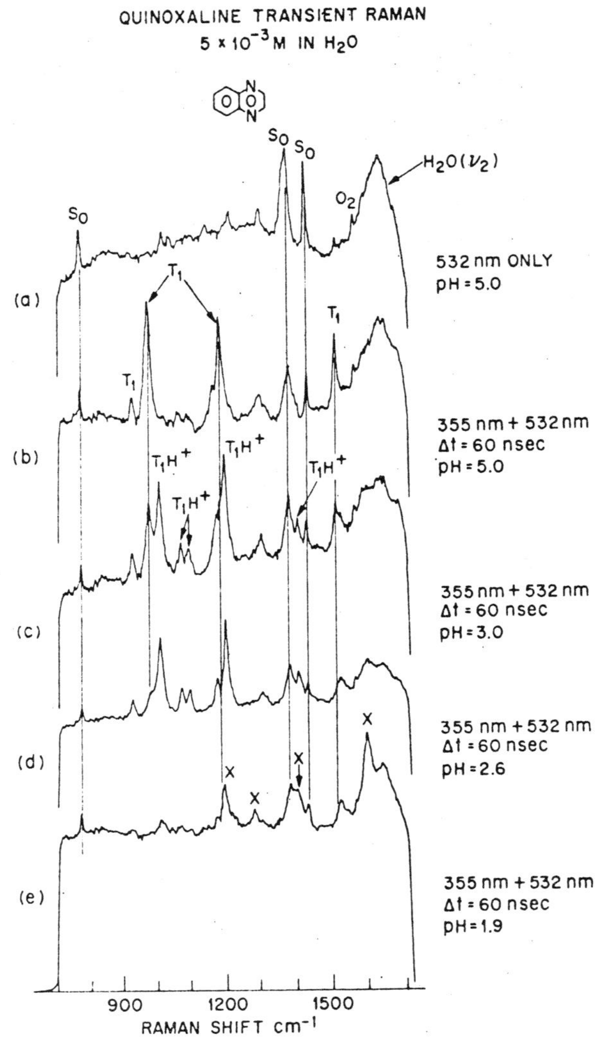

FIGURE 8 (b)-(e) transient Raman spectra of quinoxaline observed using a probe wavelength of 532 nm at a time delay of 60 ns following excitation at 355 nm as a function of pH and (a) the ground state spectrum obtained using the probe laser alone. Bands labelled T_1 are assigned to the triplet state. (Beck and Brus[25]).

typical spectrum is shown in Figure 8, from which it is evident that the triplet state decays into the protonated form, T_1H^+. Below pH 2.5, T_1H^+ evolves into a species, denoted in Figures 8 as X, which could conceivably be the doubly protonated triplet state $T_1H_2^{2+}$.

Following photoexcitation of benzoquinone, strong transient resonance Raman signals were observed from the triplet state,[26,27,28] and the semiquinone radical and radical anion.[26,27] The principal assignments are shown in Table 7.[27] These methods were used to study the photo-oxidation of water, hydrogen atom abstraction from methanol, and semiquinone-radical anion acid-base equilibria. A more recent reinvestigation of the triplet

TABLE VII Major transitions (cm^{-1}) in triplet benzoquinone and species derived therefrom.[27] Numbers in parentheses are relative intensities

S_o	T_1	$Q^{\overline{\cdot}}$	QH·	QD·
			2094 (2)	2094 (2)
	1990 (2)		1984 (1)	1984 (2)
1671 (10)	1552 (10)	1622 (10)	1619 (10)	1615 (10)
	1497 (1)	1438 (2)	1515 (6)	1515 (7)
1163 (5)	1163 (4)	1163 (4)	1163 (3)	1163 (3)
		970 (0.5)	976 (0.5)	

state of benzoquinone has given the refined values for vibrational frequencies shown in Table 8.[28] It is notable that the symmetric carbonyl stretch at 1496 cm^{-1} is usually high for $n\pi^*$ excited state, whereas the C=C stretch at 1552 cm^{-1} is low for a quinoid species. The results indicate that there is π^* antibonding density on

TABLE VIII 416 nm resonance Raman spectra (cm^{-1}) of triplet benzoquinone.[28] Numbers in parentheses are relative peak heights. The S_0 data for comparison refer to single crystals.

S_0, h_4	h_4	d_4	mode
445	455 (7)	453	CCC 6a
	901 (0.7)		
	947 (0.7)		
1160	1163 (6)	843	C-H wag 9a
	1360 (0.5)		
1660	1496 (1)	shoulder?	sym C=Q
1674	1554 (9)	1529	C=C 8a
	2003	1976	

the ring as well as the carbonyl groups.

A similar study of the anthraquinone triplet state and several species derived from this is to be found elsewhere in this volume.[29,30]

4. UNREACTIVE STATES

An example of the vibrational spectrum of the triplet state of chrysene was given earlier,[2] and further studies have been reported on this molecule using TR^3 methods,[31] which can be compared with results from coherent anti-Stokes Raman techniques.[32,33] One of the earliest observations of the vibrational spectrum of an excited state was that of the excited triplet state of p-terphenyl.[34]

TABLE IX S_o and T_1 fundamental (A_g) frequencies (cm^{-1}) in anthracene-h_{10} and anthracene-d_{10}.[35]

Approximate mode	Anthracene-h_{10}			Anthracene-d_{10}		
	S_o	T_1	$S_o - T_1$ difference	S_o	T_1	$S_o - T_1$ difference
CC stretch	1561 m	1565 w	-4	1535 m	1535 vw	-1
CC stretch	1481 w	1467 m	14	(1402) w	1348 w	61
CC stretch	1403 s	1363 s	40	1388 s	1399 s	-11
CC stretch + CH bend	1261 w	1179 m	82	1553 w	1080 m	73
CH bend	1165 m	1156 m	9	(983) vw	-	-
CH bend	1007 w	1026 w	-19	846 w	-	-
CCC torsion	745 w	740 vw	5	707 m	706 vw	1
CCC torsion	652 w	604 vw	45	601 vw	585 vw	16
CCC torsion	369 w	384 vw	-15	379 m	383 vw	-6

Vibrational assignments, shown in Table 9, based upon the TR^3 spectra shown in Figure 9[35] have been given for $^3B_{2u}{}^+$ anthracene and anthracene-d_{10}.

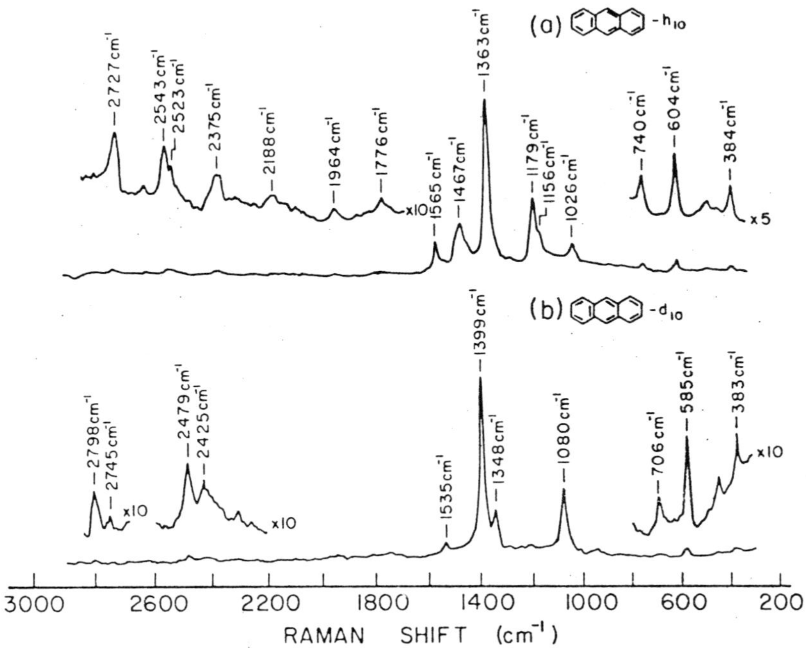

FIGURE 9 Transient resonance Raman spectra of (a) anthracene-h_{10} and (b) anthracene-d_{10} in the triplet state obtained at a probe wavelength of 427 nm, 200 ns after excitation at 337 nm. (Tripathi and Fisher[35]).

Micellar systems have been shown to increase the sensitivity in Raman spectroscopy of the T_1 states of both anthracene and chrysene.[36]

A picosecond study of a short-lived excited state of methyl orange (Scheme 4) has been performed, with results

TABLE X Vibrational assignments for methyl orange.[37]

Bands (cm^{-1})		Assignment
ground state	excited state S_1(2 < τ < 16 ps)	
1185	1180	N_α=N_β stretch
1194	same	SO_3^- stretch
1274	same	C-N stretch
1295	same	S ring[a)]
1375	same	N' ring[b)], C-C stretch
1408	1408 br	C-N' stretch
1508 sh	1498	S ring
1607	1600	S ring
1629	1618	C=C stretch N'ring

(a) S ring is the benzenesulfonate ring.

(b) N' ring is the dimethylaminobenzene ring.

shown in Figure 10.[37] The assignments of the S_1 and ground states are given in Table 10.

There have been two CARS studies on excited states of Rhodamine molecules.[38,39]

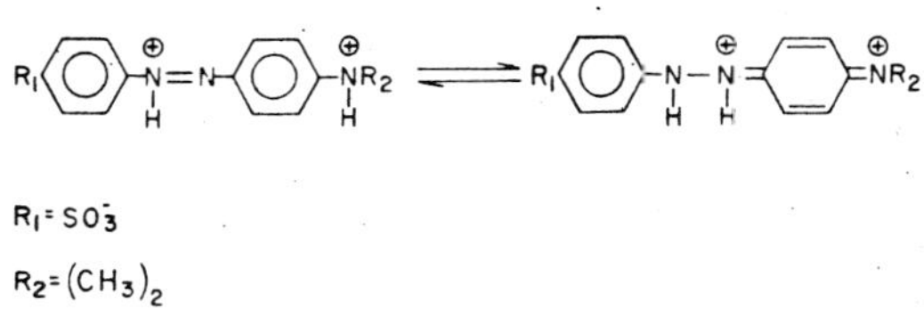

The tautomers of 4-sulfonate-4'-dimethylaminoazobenzene (methyl orange) in a strong acid environment.

Scheme 4

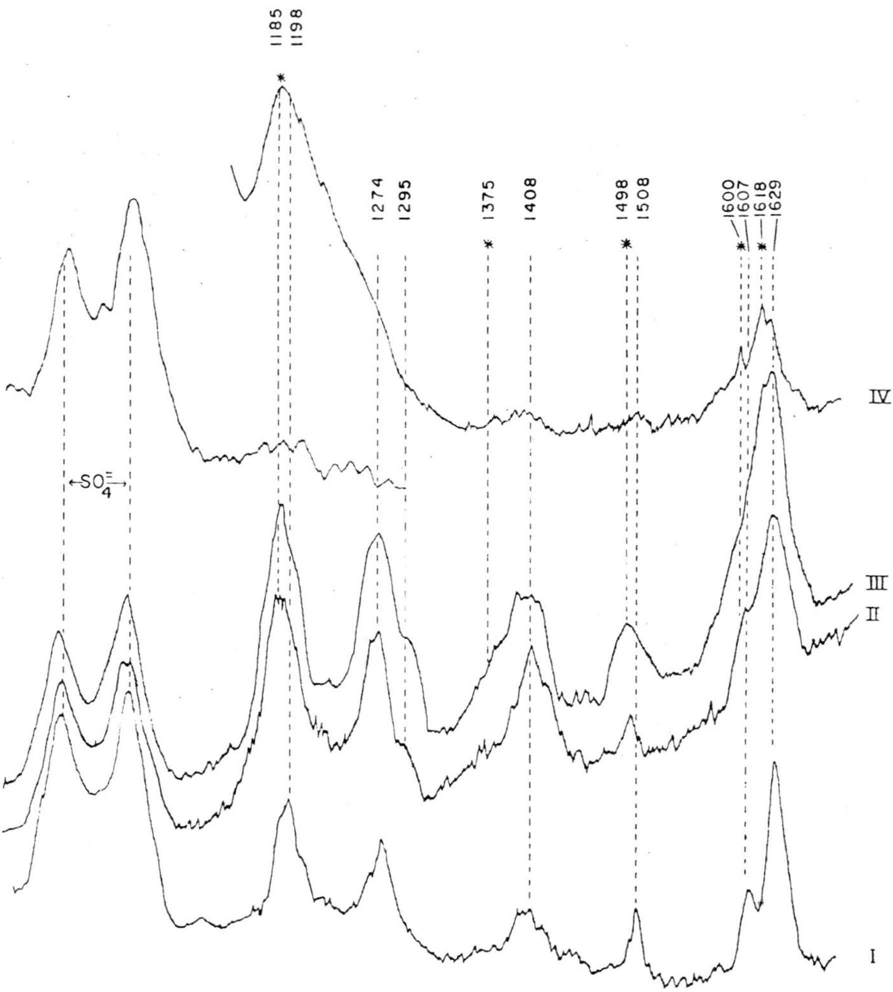

FIGURE 10

Pulsed laser Raman spectra of methyl orange in 9N H_2SO_4. (Non-resonance scattering obtained at 440.53 nm (defocused). (II) Resonance Raman scattering obtained at 490.60 nm (defocused). Ground and excited state Raman scattering of (III) 0.7 x 10^{-4} M and (IV) 1 x 10^{-5} M solutions at 490.60 nm (focused). Bands due to scattering from the S_1 state are indicated with an asterisk. (Haley and Koningstein)[37].

5. TRANSITION METAL COMPLEXES

This has been a fairly productive area of research, and it is of note that one of the earliest TR^3 spectra of an electronically excited state in solution was that of the ^3CT state of tris (2,2'-bipyridine) ruthenium (II).[40] Later work on this compound[41,42] showed that the structure of $Ru(bpy)_3^{2+*}$ is essentially that of $[Ru^{III}(bpy)_2bpy^{\bar{\cdot}}]^{2+}$. The conclusion was reached on the basis of the large frequency shifts in the MLCT (metal to ligand charge transfer) states which approximate those observed on one-electron chemical reduction of bpy to $bpy^{\bar{\cdot}}$, the TR^3 spectrum obtained on saturation of the MLCT state, which exhibits peaks attributable to both radical-like and neutral bpy, to the frequencies of 'unshifted' bpy modes in the MLCT state which resemble closely $Ru^{III}(bpy)_3^{3+}$, and also the fact that the frequency shifts in the MLCT states of the $Ru^{II}(bpy)_2^{2+}$ complexes are similar to those of the tris chelate. Some reactions of $Ru(bpy)_3^{2+}$ and proflavine with methyl viologen have also been studied with resonance Raman methods.[43]

A more recent thorough study[42,44] on the osmium complex $Os(bpy)_n(P_2)_{3-m}^{2+}$, [where P_2=cis$(C_6H_5)_2$ PCH = CHP (C H) ; n = 1-3] has confirmed that charge-transfer electron density is localised in the lowest π* orbital of one bpy ligand rather than being delocalised over the π* orbitals of all bpy ligands. Resonance Raman spectra of the $^1A_{1g}$ and $E_u(^3A_{2u})$ states of tetrakis(1,3-di-isocyanopropane) dirhodium (I),[45] and excited states of tris (2,2'-bipyridine) Cr(III)[46,47] complexes have been reported. The appropriate energy level diagram in the latter case is shown in Figure 11, and the frequencies and assignments offered are given in Table 11.

TABLE XI Frequencies of $Cr(bpy)_3{}^{3+}$ in ground ($^4A_2$) and excited ^2E and $^4T_2$ states.[47]

		$1041\ cm^{-1}$	$1326\ cm^{-1}$	$1505\ cm^{-1}$	$1573\ cm^{-1}$	$1610\ cm^{-1}$
Defocused ($^4A_2$)	FWHM (cm^{-1})	8.5 ± 0.5	8.8 ± 0.5	8.0 ± 0.5	9.5 ± 1.0	8.0 ± 0.5
cw focused (^2E)	Shifts from Def. case (cm^{-1})	-1.2	-1.2	-0.8	-1.5	-2.0
	FWHM (cm^{-1})	8.5 ± 0.5	8.6 ± 0.5	8.0 ± 0.5	7.5 ± 1.0	6.5 ± 0.5
Pulsed focused ($^4T_2$)	Shifts from Def. case (cm^{-1})	-2.5	-0.6	-2.8	-2.0	0.0
	FWHM (cm^{-1})	10.0 ± 0.5	11.0 ± 0.5	13.5 ± 0.5	11.5 ± 1.0	9.5 ± 0.5
Symmetry species in D_3		A_1	A_1	A_1	A_1	A_1
$\rho\,^4A_2$; bpy		0.2 ∿ 0.4	0.2 ∿ 0.4	0.2 ∿ 0.4	0.2 ∿ 0.4	0.2 ∿ 0.4
Assignment		ring-ring structure	CH deformation	CH deformation	in plane ring	in plane ring

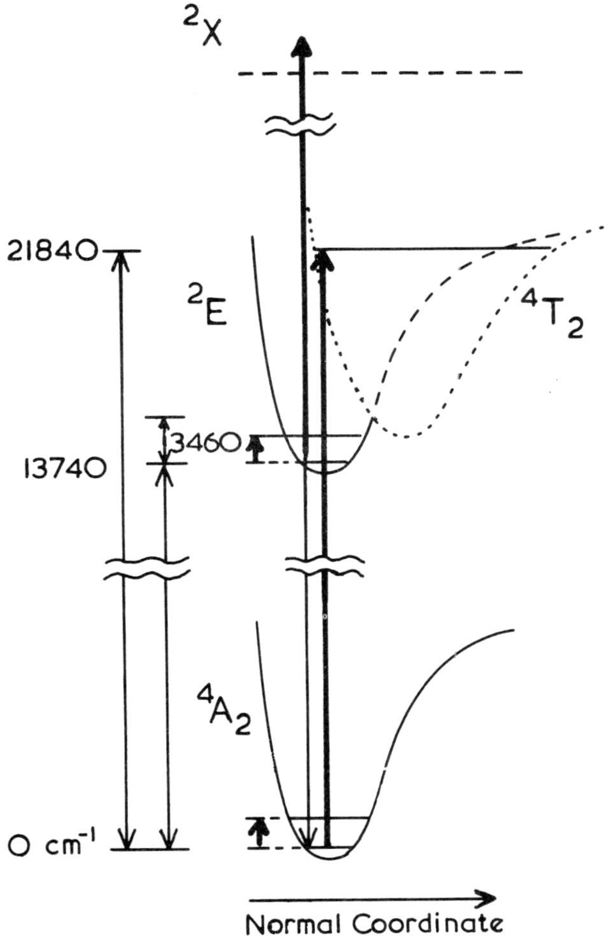

FIGURE 11 Energy level diagram of $Cr(bpy)_3^{3+}$. (Asano et al[47]).

Some early TR^3 studies in the excited ^2E state of ruby, i.e. α-Al_2O_3:Cr^{3+} should be recorded here.[48-50]

One of the most interesting of metal complexes from a biological standpoint is of course the haemoglobin molecule. An early observation of the resonance Raman spectrum of cytochrome C and oxyhaemoglobin[51] was followed by the first time-resolved measurements of carboxyhaemoglobin on the picosecond timescale,[52] illustrated in Figure 12.

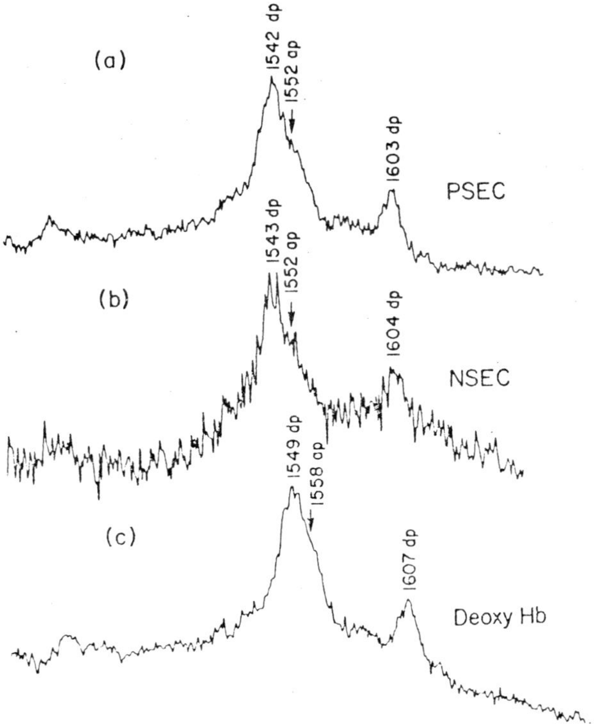

FIGURE 12 Transient resonance Raman spectra of the phototransient of caroboxyhaemoglobin obtained with pulsewidths in (a) the ps and (b) the ns time regime. (c) steady-state resonance Raman spectrum of deoxyhaemoglobin. (Terner et al[52]).

Further low-frequency observations were made,[53] and the study on carboxyhaemoglobin extended, with the results shown in Figure 13.[54] The bands due to the photoproduct resulting from photolysis of both carboxyhaemoglobin[55] and oxyhaemoglobin[56] were also observed using 30 ps pulsed excitation, bands being tabulated in Table 12. In the former case, the iron-imadazole stretching frequency was shifted to higher frequency relative to that of deoxyhaemoglobin. On the timescale used, the spectra are those of species observed before the R-T quaternary structure switch of deoxyhaemoglobin. The phorphyrin core-size bands remained at the same frequencies as for deoxyhaemoglobin, indicating that shifts observed earlier were associated with tertiary, rather than quaternary structures, and do not therefore reflect the R-T transition.

TABLE XII Resonance Raman band frequencies (cm^{-1}).[56]

	ν_{10}	ν_{19}	ν_{11}
deoxyHb	1605	1556	1549
HbCO photoproduct (and delayed HbO_2 photoproduct)	1603	1552	1543
HbO_2 photoproduct (prompt)	1590	1550	1538

6. CONCLUSIONS

The increasing appreciation of the usefulness in principle of obtaining vibrational spectra of transient electronically excited states of molecules (and species derived therefrom) will no doubt result in this area of research growing rapidly in the next few years. The extension of these methods to the picosecond time domain are already being explored. The difficulties in interpretation of time-resolved resonance Raman spectra in terms of structure

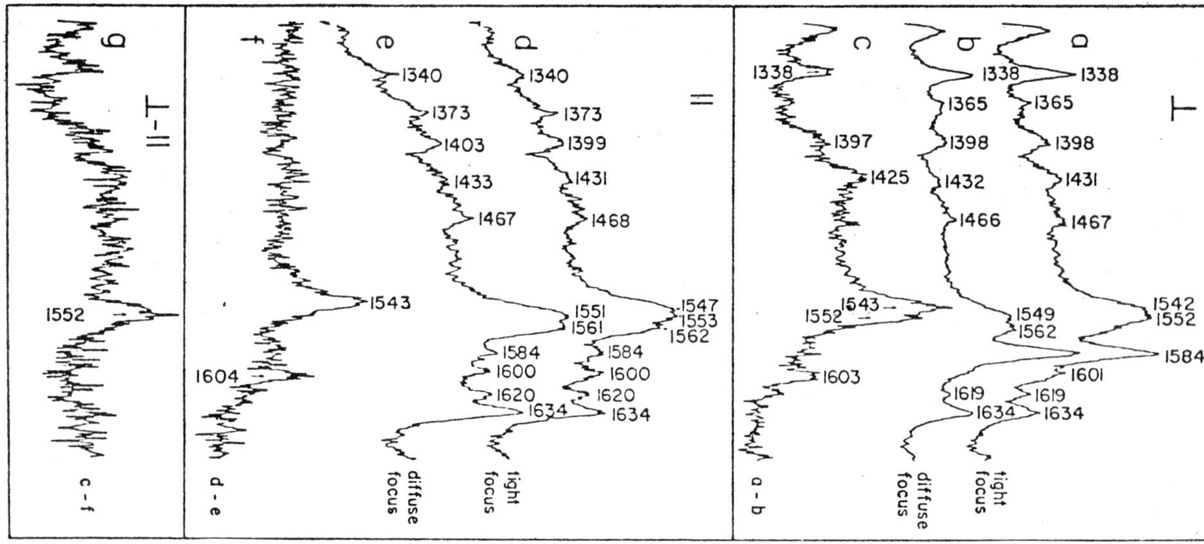

FIGURE 13 Resonance Raman spectra of carboxyhaemoglobin in perpendicular and parallel polarization. The spectra of the phototransient (c,f) are obtained by subtraction of the diffuse focus spectra (unphotolysed sample) from the tight focus spectra (partially photolysed sample). (Terner et al[54]).

of intermediates should not however be underestimated, since vibrational assignments are often difficult, and force-field calculations on excited electronic states are both absent, and fraught with difficulties. Nonetheless, the vastly improved signature of the vibrational spectrum of an excited state species when compared with the transient electronic absorption spectrum is of immense value in identifying transients, and even qualitative information about bond order changes, sites of reaction etc. is of great practical utility.

7. ACKNOWLEDGEMENTS

We are grateful to the SERC for financial support.

REFERENCES

1. G.H. ATKINSON, 'Time-resolved Raman spectroscopy' in Advances in IR and Raman spectroscopy, Vol. 9, Edited by R.E. Hester and R.J.H. Clark, Heyden, 1982.
2. G.H. ATKINSON and L.R. DOSSER, J. Chem. Phys., 72, 2195 (1980).
3. G.H. ATKINSON, 'Time-resolved Raman spectroscopy' in Advances in Laser spectroscopy, Vol. 1, Edited by B.A. Garetz and J.R. Lombard, Heyden, 1982.
4. H. HAMAGUCHI, C. KATO and M. TASUMI, Chem. Phys. Letters, 100, 3 (1983).
5. H. HAMAGUCHI, T. URANO and M. TASUMI, Chem. Phys. Letters, 106, 153 (1984).
6. T.L. GUSTAFSON, D.M. ROBERTS and D.A. CHERNOFF, J. Chem. Phys., 79, 1559 (1983).
7. T.L. GUSTAFSON, D.M. ROBERTS and D.A. CHERNOFF, J. Chem. Phys., 81, 3438 (1984).
8. R. WILBRANDT, W.E.L. GROSSMAN, P.M. KILLOUGH, J.E. BENNETT and R.E. HESTER, J. Phys. Chem., 88, 5964 (1984).
9. H. HAMAGUCHI, H. OKAMOTO and M. TASUMI, Chem. Letters 549 (1984).
10. H. HAMAGUCHI, H. OKAMOTO, M. TASUMI, Y. MUKAI and Y. KOYAMA, Chem. Phys. Letters, 107, 355 (1984).

11. R. WILBRANDT and N.H. JENSEN, J. Am. Chem. Soc., 103, 1036 (1981).
12. G.H. ATKINSON, J.B. PALLIX, T.B. FREEDMAN, D.A. GILMORE and R. WILBRANDT, J. Am. Chem. Soc., 103, 5069 (1981).
13. R. WILBRANDT, N.H. JENSEN and C. HOUEE-LEVIN, Photochem. Photobiol., 41, 175 (1985).
14. R.F. DALLINGER, J.J. GUANCI JR, W.H. WOODRUFF and M.A.J. RODGERS, J. Am. Chem. Soc., 101, 1355 (1979).
15. N.H. JENSEN, R. WILBRANDT, P.B. PAGSBERG, A.H. SILLESEN and K.B. HANSEN, J. Am. Chem. Soc., 102, 7441 (1980).
16. R.F. DALLINGER, S. FARQUHARSON, W.H. WOODRUFF and M.A.J. RODGERS, J. Am. Chem. Soc., 103, 7433 (1981).
17. L.V. HALEY and J.A. KONINGSTEIN, J. Phys. Chem., 87, 621 (1983).
18. L.V. HALEY and J.A. KONINGSTEIN, Chem. Phys., 77, 1 (1983).
19. I.W. WYLIE and J.A. KONINGSTEIN, J. Phys. Chem., 88, 2950 (1984).
20. G. HAYWARD, W. CARLSEN, A. SIEGMAN and L. STRYER, Science, 211, 942 (1981).
21. C-L. Hsieh, M. Nagumo, M. Nicol and M.A. El-Sayed, J. Phys. Chem., 85, 2714 (1981).
22. M.A. El-Sayed and J. Terner, Photochem. Photobiol., 30, 125 (1979).
23. C-L. Hsieh, M.A. EL-SAYED, M. NICOL, M. NAGUMO and J-H. LEE, Photochem. Photobiol., 38, 83 (1983).
24. C. KATO, H. HAMAGUCHI and M. TASUMI, J. Phys. Chem., 89, 407 (1985).
25. S.M. BECK and L.E. BRUS, J. Chem. Phys., 75, 4934 (1981).
26. S.M. BECK and L.E. BRUS, J. Am. Chem. Soc., 104, 1103 (1982).
27. S.M. BECK and L.E. BRUS, J. Am. Chem. Soc., 104, 4789 (1982).
28. R. ROSSETTI, S.M. BECK and L.E. BRUS, J. Phys. Chem., 87, 3058 (1983).
29. J.N. MOORE, G.H. ATKINSON, D. PHILLIPS, P.M. KILLOUGH and R.E. HESTER, 'Time-resolved resonance Raman spectroscopy of sulphonated anthraquinone derivatives', (this volume).
30. J.N. MOORE, G.H. ATKINSON, D. PHILLIPS, P.M. KILLOUGH and R.E. HESTER, Chem. Phys. Letters, 107, 381 (1984).
31. G.H. ATKINSON, D.A. GILMORE, L.R. DOSSER and J.B. PALLIX, J. Phys. Chem., 86, 2305 (1982).

32. W. WERNCKE, H.J. WEIGMANN, J. PATZOLD, A. LAU, K. LENZ and M. PFEIFFER, Chem. Phys. Letters, 61, 105 (1979).
33. H.J. WEIGMANN, K. LENZ, A. LAU, M. PFEIFFER and W. WERNCKE, Chem. Phys. Letters, 73, 175 (1980).
34. R. WILBRANDT, N.H. JENSEN, P. PAGSBERG, A.H. SILLESEN and K.B. HANSEN, Nature, 276, 167 (1978).
35. G.N.R. TRIPATHI and M.R. FISHER, Chem. Phys. Letters, 104, 297 (1984).
36. S.M. BECK and L.E. BRUS, J. Chem. Phys., 75, 1031 (1981).
37. L.V. HALEY and J.A. KONINGSTEIN, Chem. Phys., 73, 263 (1982).
38. J.P. DEVLIN and M.G. ROCKLEY, Chem. Phys. Letters, 56, 608 (1978).
39. R. KONIG, A. LAU and H.J. WEIGMANN, Chem. Phys. Letters, 69, 87 (1980).
40. R.F. DALLINGER and W.H. WOODRUFF, J. Am. Chem. Soc., 101, 4391 (1979).
41. M. FORSTER and R.E. HESTER, Chem. Phys. Letters, 81, 42 (1981).
42. P.G. BRADLEY, N. KRESS, B.A. HORNBERGER, R.F. DALLINGER and W.H. WOODRUFF, J. Am. Chem. Soc., 103, 7441 (1981).
43. M. FORSTER and R.E. HESTER, Chem. Phys. Letters, 85, 287 (1982).
44. J.V. CASPAR, T.D. WESTMORELAND, G.H. ALLEN, P.G. BRADLEY, T.J. MEYER and W.H. WOODRUFF, J. Am. Chem. Soc., 106, 3492 (1984).
45. R.F. DALLINGER, V.M. MISKOWSKI, H.B. GRAY and W.H. WOODRUFF, J. Am. Chem. Soc., 103, 1595 (1981).
46. M. ASANO, D. MONGEAU, D. NICOLLIN, R. SASSEVILLE and J.A. KONINGSTEIN, Chem. Phys. Letters, 65, 293 (1979).
47. M. ASANO, J.A. KONINGSTEIN and D. NICOLLIN, J. Chem. Phys., 73, 688 (1980).
48. L.V. HALEY, B. HALPERIN and J.A. KONINGSTEIN, Chem. Phys. Letters, 54, 389 (1978).
49. B. HALPERIN and J.A. KONINGSTEIN, J. Chem. Phys., 69, 3302 (1978).
50. B. HALPERIN, D. NICOLLIN and J.A. KONINGSTEIN, Chem. Phys., 42, 277 (1979).
51. R.B. SRIVASTAVA, M.W. SCHUYLER, L.R. DOSSER, F.J. PURCELL and G.H. ATKINSON, Chem. Phys. Letters, 56, 595 (1978).

52. J. TERNER, T.G. SPIRO, M. NAGUMO, M.F. NICOL and M.A. EL-SAYED, J. Am. Chem. Soc., 102, 3238 (1980).
53. M.J. IRWIN and G.H. ATKINSON, Nature, 293, 317 (1981).
54. J. TERNER, J.D. STRONG, T.G. SPIRO, M. NAGUMO, N. NICOL and M.A. EL-SAYED, Proc. Natl. Acad. Sci. USA, 78, 1313 (1981).
55. P. STEIN, J. TERNER and T.G. SPIRO, J. Phys. Chem., 86, 168 (1982).
56. J. TERNER, D.F. VOSS, C. PADDOCK, R.B. MILES and T.G. SPIRO, J. Phys. Chem., 86, 859 (1982).

CHAPTER 4

INVERSE PRE-RESONANCE RAMAN AND COHERENT AMPLIFICATION STUDIES IN DYE SOLUTIONS

W. J. JONES
Edward Davies Chemical Laboratories,
University College of Wales,
Aberystwyth,
Dyfed SY23 1NE, Wales.

To some extent the work to be described in this paper is somewhat misplaced in a programme on Time-resolved Laser Raman Spectroscopy since it is not involved directly with the investigation of the Raman spectra of transient molecular species and indeed much of the work is not even involved with Raman spectroscopy. However it is directly relevant to Raman spectroscopy and the techniques, although comparatively new, will find wide application in Raman spectroscopy generally and have the potential to be extended readily into the investigation of transient molecular species, one of the most promising areas of investigation using the techniques employed in these studies.

The general area of investigation is Inverse Raman Spectroscopy, a particular variant of Raman Amplification spectroscopy in which the intensity loss on a higher frequency laser is recorded under the perturbing influence

c

of a lower frequency laser when the frequency difference of the two lasers coincides with the frequency of a Raman active transition in the molecular system under study.

The methods that have been developed for recording Inverse Raman and Raman Gain spectra[1] are proving to be ideal also for investigating pico-second processes generally and both aspects of our studies are considered in this paper.

The experimental arrangement employed is shown in Figure 1. A mode-locked argon laser synchronously pumps two dye lasers. The two dye laser beams are incident on the sample under study, one of the lasers operating at a fixed wavelength the other being wavelength tunable. Both lasers are picosecond lasers operating at a pulse repetition rate of 82 MHz, the duration of individual pulses being constrained to ∿15 ps in order to ensure a reasonably narrow spectral line width (∿1-2 cm^{-1}) for the transform limited laser pulses. The cross-correlation temporal profile of these lasers is typically ∿30 ps, indicating relatively little time jitter between the pulses of the two lasers. When pulses from the two lasers are precisely coincident in time conditions are optimal for recording two photon processes, in particular two photon absorption, in which one photon is absorbed from each laser, or Raman spectroscopy, in which one photon is absorbed from the higher frequency laser and one photon is created in the lower frequency laser, with the energy balance serving to excite a Raman active vibrational state.

By a modulation of one of the lasers, the pump laser, at some appropriate frequency it is possible using conventional lock in detection methods to record the appearance of intensity gain or loss on the unmodulated laser, the

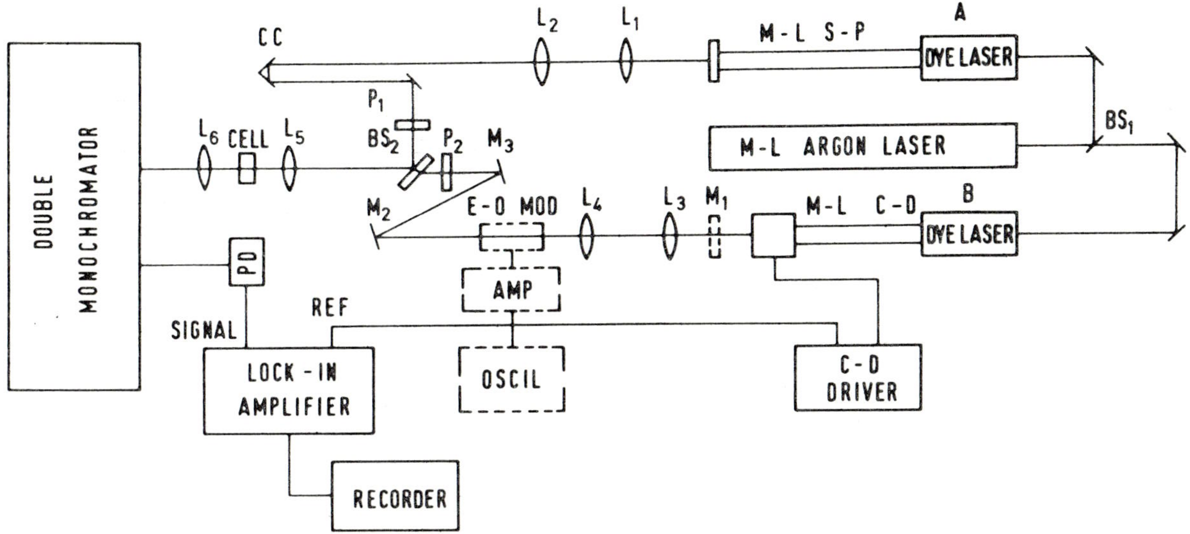

FIGURE 1 The picosecond coherent amplification spectrometer. A: mode-locked synchronously pumped dye laser; B: mode-locked synchronously pumped and cavity-dumped dye laser; L: lenses; M: mirrors; BS: beam splitter; P: polarizer; CC: distance scannable corner cube retro reflector; PD: photodiode [published with permission from J. Chem. Soc.; Faraday Transactions II, 1983, 79, 865].

probe laser, as a result of the Raman mixing process. With approximately 2 mW of probe laser power falling on a high quantum efficiency silicon detector the minimum extractable signal at the short noise limit corresponds to a modulated loss or gain in the intensity of the probe laser of $\sim 10^{-8}$. By scanning the wavelength of one or other of the lasers it is possible to monitor the loss or gain of the other laser as a function of the sum wavenumber of the two lasers, the latter condition for the study of Raman spectra the former for two photon absorption spectroscopy.

Figure 2 shows the spectrum obtained when using Rhodamine 6G dye in solution in methanol at a concentration of 10^{-3} mol ℓ^{-1}. This spectrum[1] shows some very interesting features which formed the basis for a fairly detailed study of the effects produced when two picosecond lasers interact in a dye solution. This spectrum shows primarily the Resonance Inverse Raman spectrum of Rhodamine 6G. Additionally, as might have been expected, it shows evidence of a non-resonantly enhanced band due to the solvent, methanol, but also it shows the appearance of a pronounced feature at low wavenumber displacements of one laser from the other exhibiting probe laser gain rather than loss. This gain process arises not from a two photon process but from two consecutive one phoon processes when the lasers lie in the vicinity of the dye absorption band. In essence the pump laser excites molecules in the ground electronic state S_0 into the excited electronic state S_1, the concentration of molecules in S_1 being interrogated by the probe laser via the stimulated emission process or, alternatively, by locating the probe laser in the region of sample absorption rather than

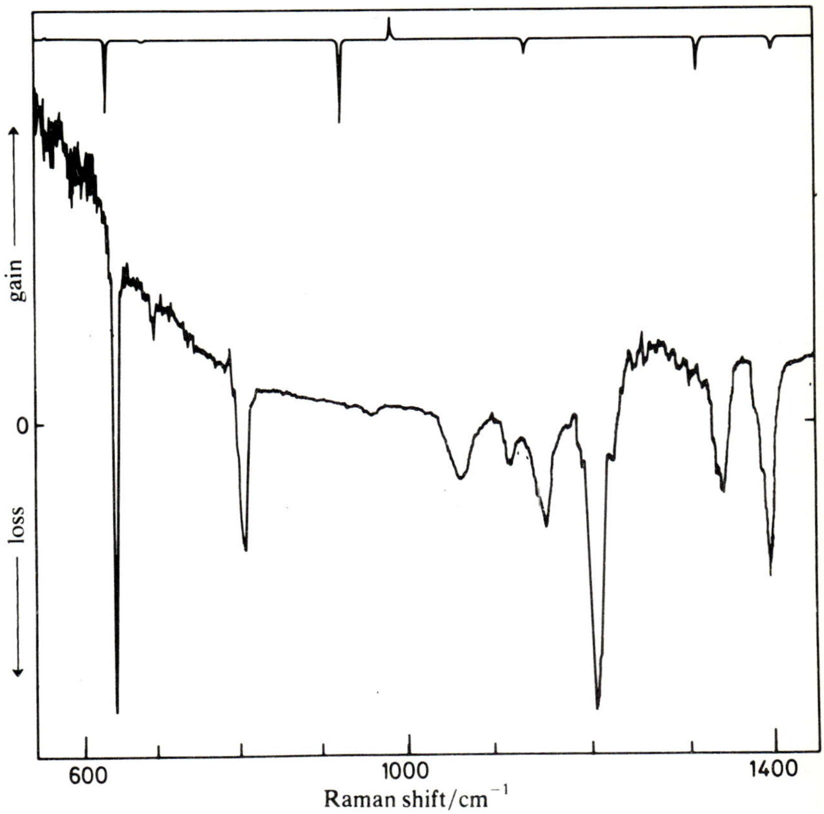

FIGURE 2 The inverse pre-resonance Raman spectrum of 10^{-3} mol ℓ^{-1} Rhodamine 6G dye in methanol solution. For this spectrum the probe laser was set at 614 nm. The trace above the spectrum displays the opto galvanic calibration spectrum of an iron-neon hollow-cathode lamp. [Published with permission from J. Chem. Soc., Faraday Transactions II, 1983, 79, 865].

sample fluorescence it is possible to probe the "hole" in the ground state concentration produced as a result of absorption of pump laser photons. Since the lasers can be scanned in time relative to one another it is possible to derive information on orientational and lifetime decay processes because an ordered array of excited electronic state or ground state species can be prepared by one polarized laser and interrogated quantitatively by the other. It is interesting to note that in the spectrum of Rhodamine 6G the lasers are displaced by $\sim$2800 cm^{-1} from the electronic absorption band origin and at this wavelength it is extremely difficult to monitor dye absorption for the cell length employed here using conventional spectrophotometric methods. By contrast these coherent amplification methods generate a gain of $\sim 10^{-4}$ and a realizable signal to noise ratio of $\sim 10^3$-10^4. The methods developed for investigating two-photon Raman processes have provided an additional benefit in this ability to record conventional absorption spectra at a sensitivity substantially greater than can be obtained by conventional spectrophotometry, potentially an enormous boon for investigating ultra thin layers of absorbing species.

Since the two lasers are focussed together into the sample the excitation volumes are small, typically in the region of 20 picolitres or less, with minimum detectable Raman signals from $\sim 10^9$-10^{10} molecules, Resonance Raman spectra from 10^5-10^6 molecules, and consecutive one photon process signals from as few as $\sim$100 molecules of an absorbing species.

Figure 3 shows the absorption/fluorescence bands of cresyl violet with absorption maximum at 600 nm and fluorescence maximum at 628 nm. Since the fluorescence gain

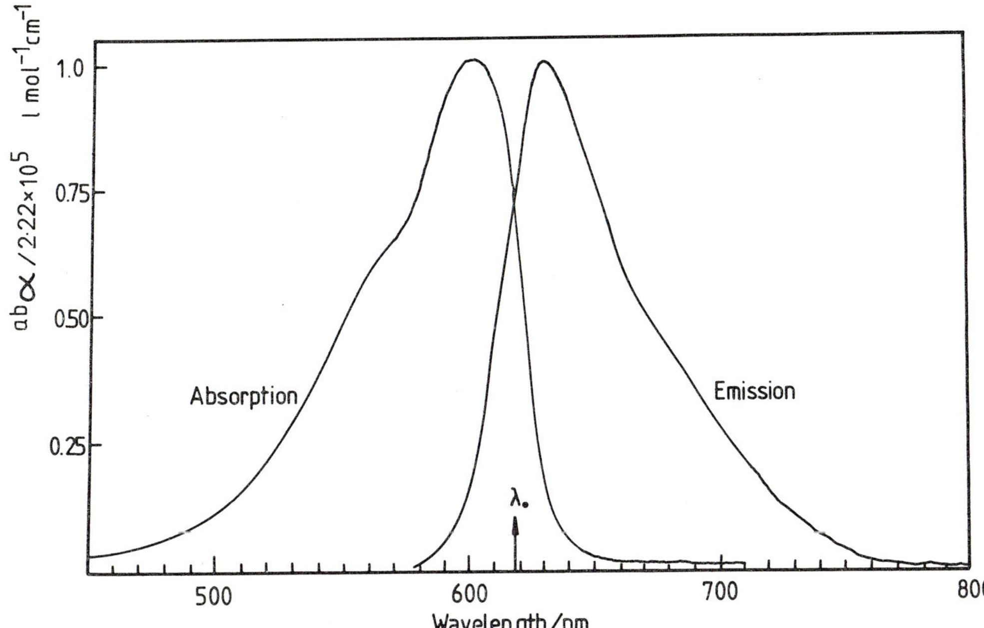

FIGURE 3 The absorption spectrum of cresyl violet in ethanol at a concentration of 2.5×10^{-6} mol ℓ^{-1} in a 1 cm cell. The emission curve normalised to the peak of the absorption curve is also shown. [Published with permission from Chemical Physics].

arises from two consecutive single photon processes it should be quantifiable from data derived from conventional spectrophotometric methods. From a detailed consideration of the absorption-stimulated emission processes occurring in the dye medium we derive an expression for the probe laser gain[2] of

$$G^{(\ ,\ ),t} = F^{(\ ,\ ),t} \,.\, C \,.\, {}^{ab}\alpha_{p}\left({}^{ab}\alpha_{pr} + {}^{se}\alpha_{pr}\right).\frac{{}^{i}p\,\ell^{2}}{LV}, \tag{1}$$

where G is the calculated gain, C the sample concentration, ${}^{ab}\alpha_{p}$ the absorption coefficient at the pump laser wavelength, ${}^{ab}\alpha_{pr}$ and ${}^{se}\alpha_{pr}$ the corresponding coefficients for absorption and stimulated emission at the wavelength of the probe laser. ${}^{i}p$ is the number of pump laser photons in each pulse, ℓ the effective length of the focal region and V the effective focal volume, L being Avogadro's constant. Since we are dealing with focussed Gaussian beams ℓ and V are inter-related and this expression can be replaced by an experimentally more useful form[2]

$$G^{(\ ,\ ),t} = 1.5 \times 10^{10}.\ A.\ F^{(\ ,\ ),t}\left({}^{ab}\alpha_{pr} + {}^{se}\alpha_{pr}\right). \frac{{}^{i}p}{L\lambda}.\ \ell n\left(\frac{1}{T_{p}}\right), \tag{2}$$

where T_p is the transmission of a 1 cm sample of the solution under study, λ the probe wavelength and A a constant that takes account of the fact that only 50% of the total signal is generated in the confocal region of the focussed beams. A is expected to be very nearly equal to 2 and quantitative comparisons of this expression with the experimental data do indeed show excellent agreement for A $\sim$1.2-1.4. Such a comparison is shown in Figure 4, the solid line being the measured gain as a function of pump laser wavelength. The circles on this diagram are the calculated gain values after making due allowance for the variation of pump laser intensity with wavelength. The agreement is seen to be excellent and confirms the essentially linear nature of the consecutive single photon processes.

The Gain coefficient, G, is a function of the time delay, t, between the pump and probe lasers and also depends on the relative polarizations of the two lasers, for parallel polarizations or   for perpendicular polarizations of the two lasers. The term F demonstrates how the gain depends on these polarizations as a function of time. Since the pump laser is plane polarized and excites only those molecules whose transition dipoles are oriented appropriately in relation to this polarization, an ordered array of molecular dipoles is produced. The probe laser following at a later time investigates the rate at which this ordered array both relaxes back to a perfectly random distribution and loses it excitation energy to revert to the molecular electronic ground state.

By scanning the time difference between the two lasers it is possible to derive temporal decay curves for the parallel or perpendicular polarizations of the two lasers at any pump and probe laser wavelengths. The parallel case depends on $F^{11,t}$ given by[2]

$$F^{11,t} = [1 + \frac{4}{5}\exp(-\frac{t}{\tau_{or}})]\exp(-\frac{t}{\tau_F}), \qquad (3)$$

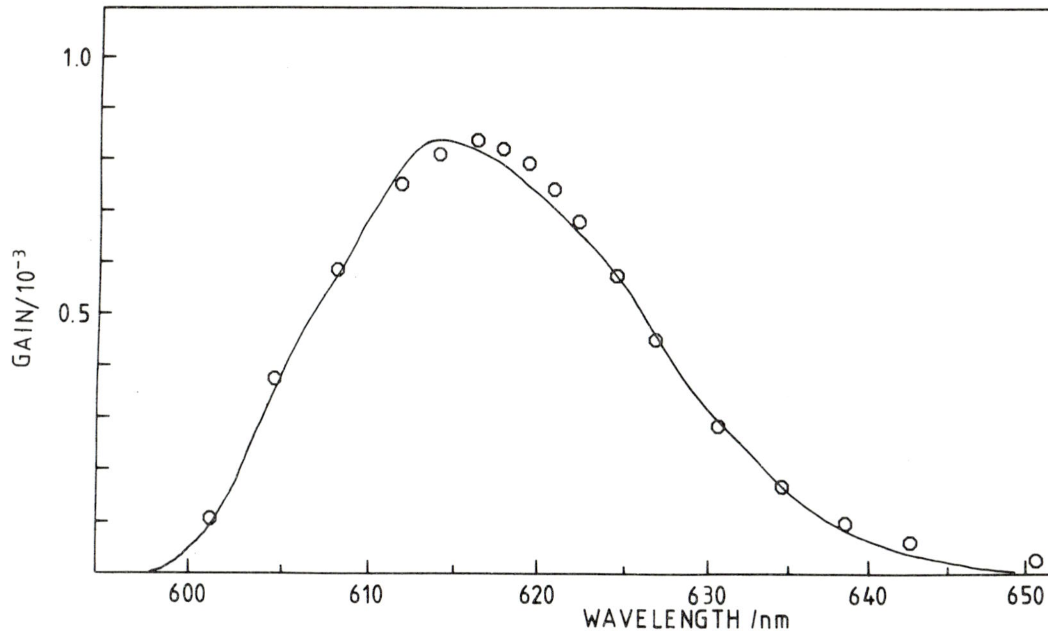

FIGURE 4 The gain spectrum of cresyl violet. The wavelength of the probe laser was set at 588 nm. The open circles are calculated from equation (2) with $A = 1.3_3$. [Published with permission from Chemical Physics].

and the perpendicular case by

$$F^{\perp,t} = [1 - \frac{2}{5}\exp(-\frac{t}{\tau_{or}})]\exp(-\frac{t}{\tau_F}), \qquad (4)$$

where τ_{or} is the decay time for molecular reorientation and τ_F the lifetime of the fluorescing state. These decay curves are shown in Figure 5.

It is possible to obtain τ_{or} and τ_F by appropriate computer simulation of the decay curves represented by equations (3) and (4) and displayed in Figure 5 but the method is tedious and subject to experimental uncertainties. A more satisfactory method is derived by appropriate polarization combinations of these expressions which can be performed experimentally yielding decay curves in which τ_{or} and τ_F dominate the experimental decay processes.

The appropriate fluorescence decay laws are:

(1) For an intensity modulated pump beam whose polarization direction is oriented at an angle of 54.7° with respect to the probe beam

$$F^{54.7°,t} = \exp(-\frac{t}{\tau_F}) \qquad (5)$$

(2) For a polarization modulated pump beam

$$F^{pol,t} = \exp(-t(\frac{1}{\tau_{or}} + \frac{1}{\tau_F})) \qquad (6)$$

Decay curves appropriate to the polarization combinations are shown in Figure 5, these curves yielding τ_{or} = 104 ps for the orientational decay of cresyl violet in methanol and τ_F = 3.6 ns for the lifetime decay.

These methods are generally applicable for the

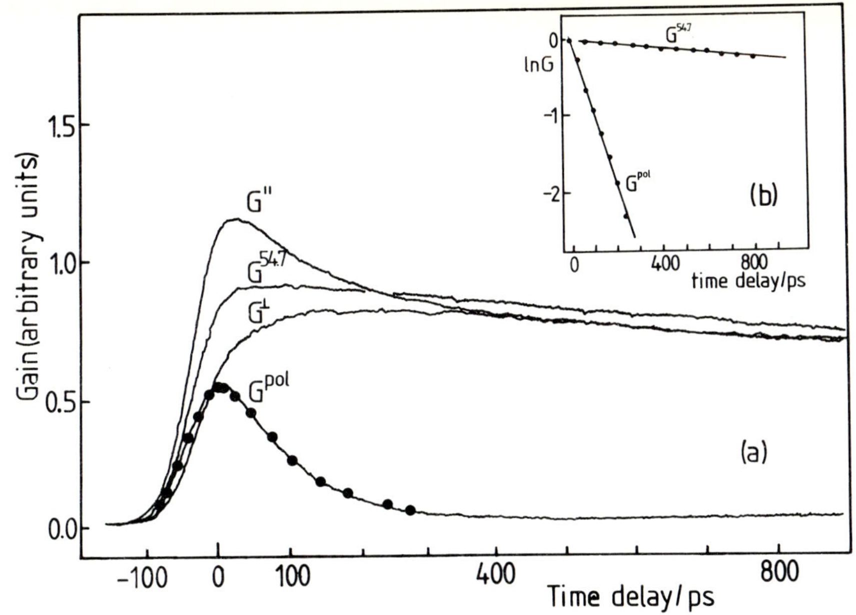

FIGURE 5 (a) Gain decay curves for cresyl violet. The decay curves are governed by expressions (2)-(6). The closed circles in this diagram are the differences ($G^{ll,t} - G^{L,t}$) at specified times and are compared with the experimental decay curve, $G^{pol,t}$.
(b) Semi-log plots of intensity modulation$^{54.7}$ and polarization modulation, G^{pol}, decay curves. [Published with permission from Chemical Physics].

investigation of molecular reorientation and lifetime decay processes, but additionally there are many aspects of these studies which have proved to be extremely interesting in relation to the study of Raman spectroscopy.

To the present most resonance Raman spectra have been recorded via Raman emission spectroscopic techniques or via Coherent Anti-stokes Raman Spectroscopy (CARS). These studies have been extended extensively to biological systems. Our studies suggest that the Inverse Raman process will prove to be a strong competitor to these techniques. As an indication of the quality of the spectra which may be obtained Figure 6 shows a series of spectra obtained at concentrations of Rhodamine 6G in the range 1×10^{-1} mol ℓ^{-1} to 4×10^{-4} mol ℓ^{-1}. For this study[3] the probe laser was set at 614 nm, some 2500 cm^{-1} from the electronic band origin. Since the spectrum is recorded in the pre-resonance region the resonance enhancement is markedly less than would be obtained by moving the probe laser toward the electronic band origin. The lines are found to be Lorentzian over the whole concentration range, unlike the situation that occurs in CARS studies. Since the loss is readily measured it is apparent that the technique lends itself to quantitative studies, most readily by comparing the measured loss with that recorded for the 1029 cm^{-1} methanol band, for which the cross section is known. By measuring the spectra for crossed polarizations of the two beams it is possible to obtain depolarization measurements as in conventional Raman emission spectroscopy.

Before completing this treatment of the Inverse Raman method I should consider one further aspect in which this method is likely to prove of very great value, I refer to

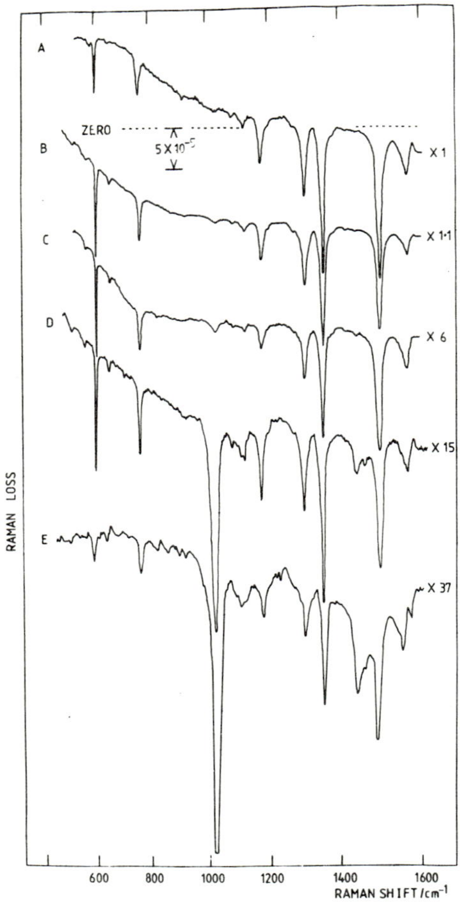

FIGURE 6 Inverse pre-resonance Raman spectra of Rhodamine 6G in solution in methanol at concentrations of: A $\sim 1\times10^{-1}$ mol ℓ^{-1}; B -4×10^{-2} mol ℓ^{-1}; C -1×10^{-2} mol ℓ^{-1}; D $\sim 2\times10^{-3}$ mol ℓ^{-1}; E $\sim 4\times10^{-4}$ mol ℓ^{-1}. The Raman loss is indicated on spectrum A, this scale being expanded by the appropriate multiplying factor for each of the other spectra. [Published with permission from Chemical Physics].

its ability to generate Raman spectra free of the effects of sample fluorescence. The analogue of sample fluorescence recorded in Raman emission studies is the fluorescence gain generated in our Inverse Raman experiments. Since our studies are of a quantitative nature it is possible to compare the relative merits of the spontaneous emission methods with Inverse Raman spectroscopy. In Raman emission studies a monochromatic laser beam generates Raman emission to longer wavelength, where fluorescence is generally concentrated. In the Inverse Raman method the pump laser lies to long wavelengths of the detection element, in this case the probe laser. The effectiveness of this method in the rejection of sample fluorescence is quite dramatic. Figure 7 shows portions of the spectra of CS_2:Ethanol solvent mixtures containing known amounts of a fluorescing dye impurity, in the example shown Rhodamine 6G.[4] The upper two traces display spontaneous emission spectra at concentrations of 1.3×10^{-5} mol ℓ^{-1} and 1.3×10^{-6} mol ℓ^{-1}, the lower two being Inverse Raman spectra at concentrations of 4×10^{-5} mol ℓ^{-1} and 4×10^{-4} mol ℓ^{-1}. The spectral regions recorded do not quite overlap although all the features recorded in the emission spectrum are present in the Inverse Raman spectrum. The locations of the laser and Raman transitions in the two cases are near identical so that the spectra are directly comparable. In the Inverse Raman spectrum it is apparent that the concentration of dye has been increased to such an extent that resonance Raman features of the dye impurity are becoming apparent.

A comparison of the Raman signal intensity to fluorescence background for both methods shows an improvement for the Inverse Raman aproach using picosecond lasers of 3

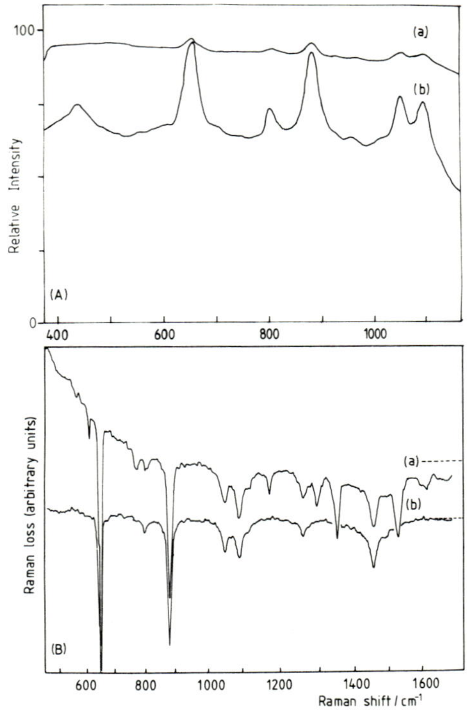

FIGURE 7 Raman emission (A) and inverse Raman (B) spectra of CS_2:ethanol solutions (1:9 ratio) in the presence of Rhodamine 6G. The concentrations of the Rhodamine 6G in (A) are: (a) $\sim 1.3\times10^{-5}$ mol ℓ^{-1}; (b) $\sim 1.3\times10^{-6}$ mol ℓ^{-1}. The concentrations of the Rhodamine 6G in (B) are: (a) $\sim 4\times10^{-4}$ mol ℓ^{-1}; (b) $\sim 4\times10^{-5}$ mol ℓ^{-1}. The zero for each inverse Raman spectrum is indicated by the dotted line. The additional lines in (B) are resonance Raman transition in Rhodamine 6G brought into the observational range as a result of the high concentration employed. [Published with permission from Chemical Physics].

orders of magnitude for transitions in the vicinity of $\sim$500-600 cm^{-1}, rising to $\sim$4 orders of magnitude for the Raman displacements in excess of $\sim$1000-1200 cm^{-1}.

The methods of Inverse Raman spectroscopy are still in their infancy but the potential for further development is extraordinarily high both for conventional and resonance enhanced spectra and, in due course, for the study of transient species.

ACKNOWLEDGEMENTS

It is a pleasure to express appreciation for financial support from the Science and Engineering Research Council, with the aid of which this work was carried out. The individuals concerned with promoting the work are Dr. A. J. Langley, Dr. J. Baran, Dr. A. Grofcsik, Mr. R. A. Beaman (SERC CASE award with ICI) and Mr. A. Davies.

REFERENCES

1. J. BARAN, D. ELLIOT, A. GROFCSIK, W. J. JONES, M. KUBINYI, A. J. LANGLEY and V. U. NAYAR, Raman Amplification Spectroscopy using Mode-locked Lasers, J. Chem. Soc., Faraday Transactions 2, 79, 865, (1983).
2. J. BARAN, A. J. LANGLEY and W. J. JONES, Coherent Amplification in Fluorscent Dye Solutions. I. The Fluorescence Gain Spectrum of Cresyl Violet, accepted for publication in Chemical Physics.
3. J. BARAN, A. J. LANGLEY, A. GROFCSIK and W. J. JONES, Coherent Amplification in Fluorescent Dye Solutions. II. The Inverse Pre-resonance Raman Spectrum of Rhodamine 6G, accepted for publication in Chemical Physics.
4. J. BARAN, R. A. BEAMAN, A. J. LANGLEY and W. J. JONES, Coherent Amplification in Fluorescent Dye Solutions. III. On the Merits of Inverse Raman Spectroscopy for the Rejection of Sample Fluorescence, accepted for publication in Chemical Physics.

CHAPTER 5

TIME-RESOLVED RESONANCE RAMAN SPECTROSCOPY OF SULPHONATED ANTHRAQUINONE DERIVATIVES

J.N. MOORE and D. PHILLIPS
Davy Faraday Research Laboratory,
The Royal Institution,
21 Albemarle Street,
London. W1X 4BS

R.E. HESTER,
Department of Chemistry,
University of York,
Heslington,
York. YO1 5DD

G.H. ATKINSON and P.M. KILLOUGH,
Department of Chemistry,
University of Arizona,
Tucson,
Arizona 85721,
U.S.A.

1. INTRODUCTION

9,10-Anthraquinone and its derivatives, particularly the sulphonated compounds, have been studied extensively because of their high photochemical activity.[1,2] The two 'strong' sensitizers anthraquinone-2-sulphonate (AQ2S) and anthraquinone-2,6-disulphonate (AQ26DS) have received much attention because of their high water solubility and wide range of photosensitizing properties, and have been used as models for various other more complex systems such as the series of anthraquinonoid vat dyes.[3-8] Studies have been reported using flash photolysis, pulse radiolysis, and steady-state irradiation techniques, recent work being stimulated by reports that sulphonated anthraquinones may be used as photosensitizers in the catalytic generation of chlorine from chloride ions[9] or in water splitting,[10] using solar energy. Although these compounds are favourable for such applications because they can be used in oxygenated aqueous solution, there is a serious disadvantage since they undergo photochemically driven reactions to produce hydroxylated derivatives as permanent products.[11-15] Despite the many studies on these compounds, it is only recently that the mechanisms by which reactions occur have become clear. All reactions studied so far have involved the triplet state, since intersystem crossing is efficient ($\emptyset > 0.9$)[16,17] and the singlet state is estimated to be short-lived.[18] The photochemistry in alcoholic solution has been understood for many years, and is known to involve hydrogen atom abstraction by the triplet state.[3,19,20] In aqueous solution the scheme is more complex, but may be divided into two situations. In the presence of high concentrations of a strong electron donor, electron transfer will occur to produce the radical anion (AQ^{-}) which will protonate in acidic conditions to generate the neutral semiquinone radical ($AQH^{\bullet}$).[24] In pure aqueous solution, up to three transients have been observed following decay of the triplet and have been assigned to two triplet-water complexes labelled B and C[23,24] and under basic conditions (pH > 11) to the radical anion ($AQ^{\bar{\bullet}}$).[24] The triplet and other species described above have been observed on a nanosecond to microsecond timescale by their optical absorption properties, and the overall scheme proposed for their reaction is complex.[24]

Although the flash photolysis experiments have provided detailed information on both the identification

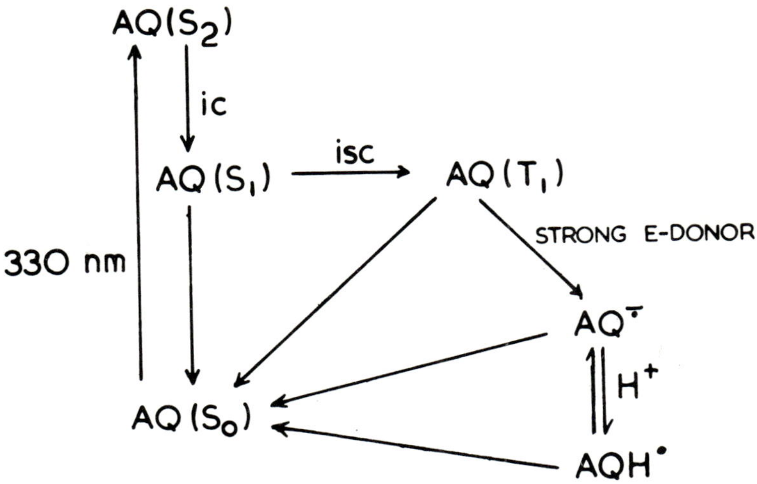

FIGURE 1 Scheme for reaction of sulphonated anthraquinone derivatives following photolysis in the presence of a strong electron donor.

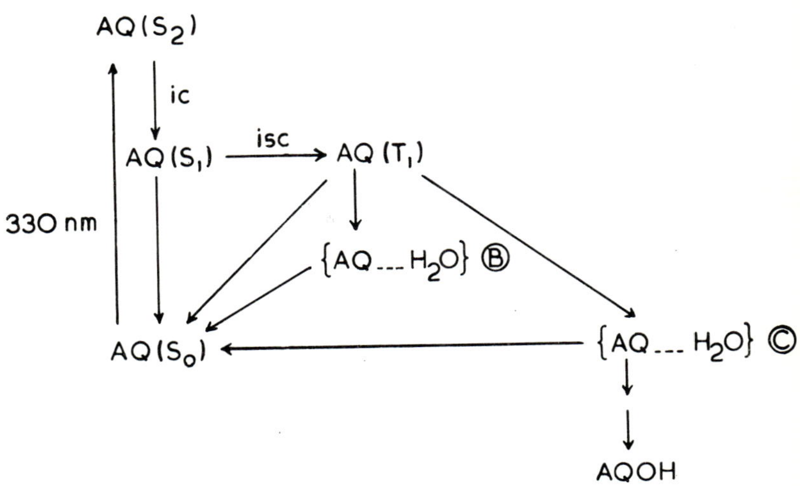

FIGURE 2 Scheme for reaction of sulphonated antraquinone derivatives following photolysis in aqueous solution.

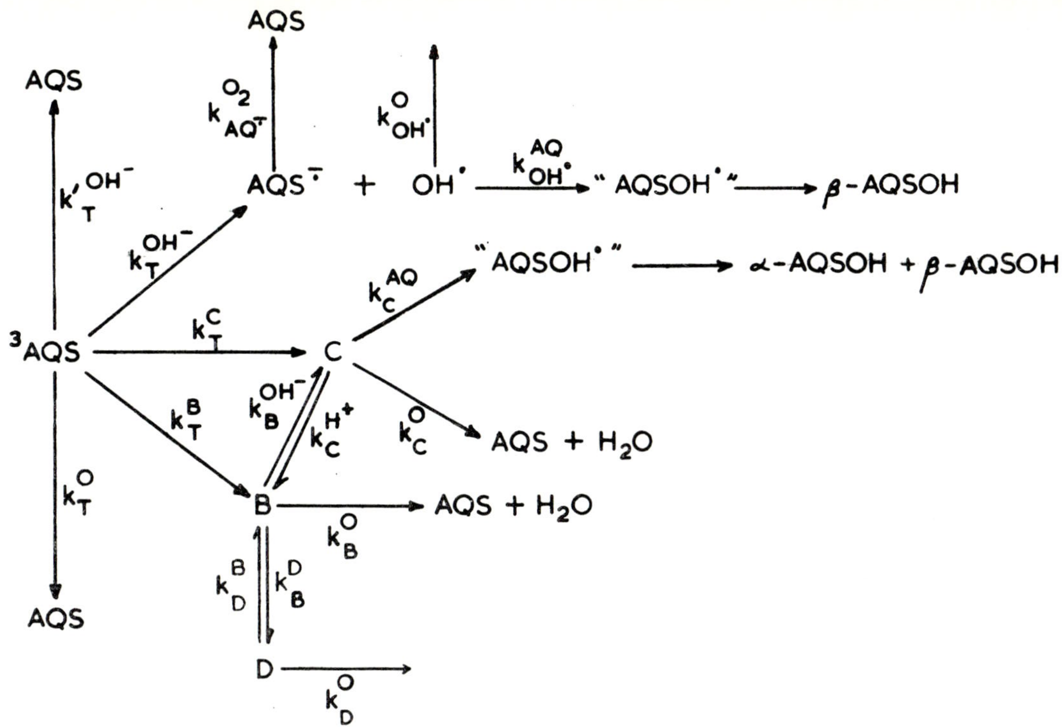

FIGURE 3 Complete scheme for reaction of the triplet state of sulphonated anthraquinone derivatives in aqueous solution.

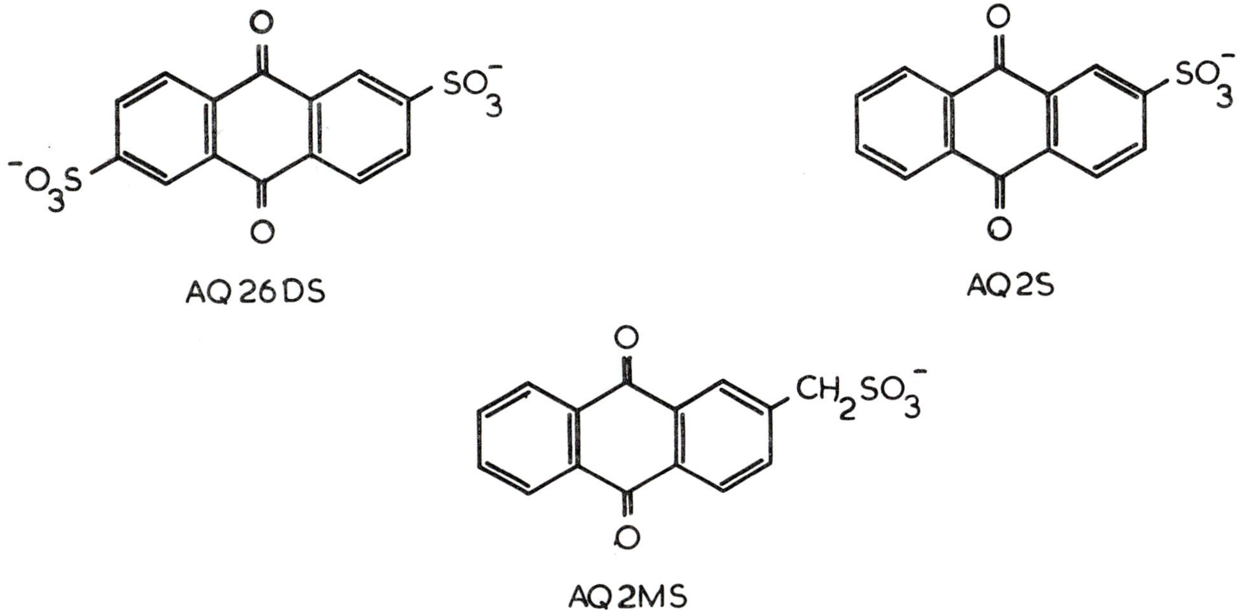

FIGURE 4 Structures of anthraquinone-2,6-disulphonate (AQ26DS), anthraquinone-2-sulphonate (AQ2S), and anthraquinone-2-methylsulphonate (AQ2MS).

of species present under various conditions and the kinetics of their formation and decay, several questions remain concerning the nature of these species. The broad absorption bands provide no information on the vibrational spectra of the transients and therefore allow only speculation on what their structures may be. In particular, knowledge of the structures of the triplet-water complexes B and C is important since any catalytic photochemical cycle would be greatly improved by the absence of such side reactions, and information on their structures gives the possibility of synthesising molecules with sites blocked, thereby preventing their formation. This would be especially desirable in the case of transient C since this species leads to production of hydroxylated derivatives and permanent loss of photosensitizer.

Time-resolved resonance Raman spectroscopy (TR^3) would seem to be the ideal technique for an investigation into the structures of such short-lived species, since the selective nature of the resonance enhancement allows the study of individual species in a complicated mixture where there are overlapping absorption bands.[25] Hence, a study of the three 'strong'sensitizers AQ2S, AQ26DS and anthraquinone-2-methyl-sulphonate (AQ2MS) using TR^3 has been initiated, and here we report on our preliminary results concerning these compounds.

2. EXPERIMENTAL

The experimental arrangement is illustrated in Figure 5. Excimer 1 (Lambda Physik EMG201) is used as the photolysis source at either 337 nm (N_2 fill, 6 mJ/pulse, 6 ns pulse-width) or 351 nm (XeF fill, 120 mJ/pulse filtered to

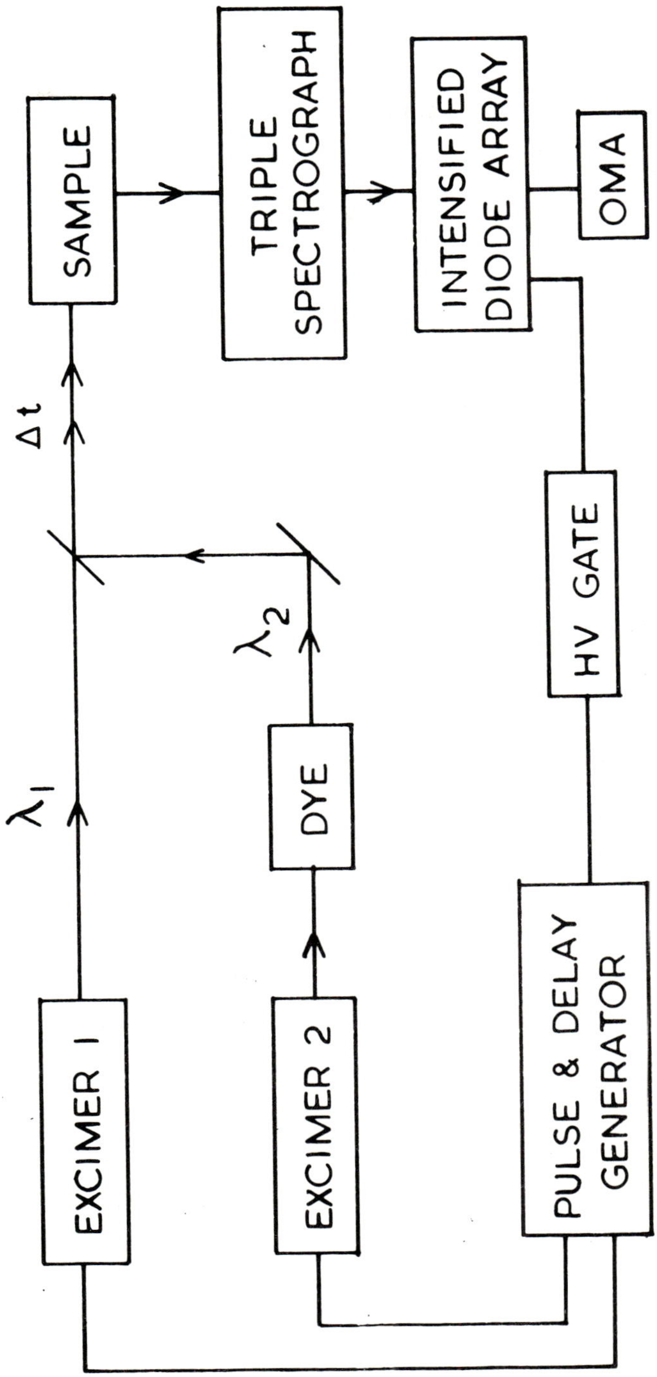

FIGURE 5 Experimental arrangement used for the TR^3 experiments.

30 mJ/pulse, 10 ns pulsewidth). Excimer 2 (Lambda Physik EMG101; XeCl fill) pumps a dye laser (Lambda Physik FL2002) which has been used at various wavelengths between 405 nm and 540 nm (5-10 mJ/pulse, 10 ns pulsewidth). The laser beams are focussed below the sample to avoid non-linear optical effects which can be encountered using the high power densities available with these lasers, and the sample is flowed in either an open system or a closed quartz capillary tube at a rate sufficient to ensure that each pulse pair encounters fresh solution. The scattering is collected at right angles to the incident beam, and analysed using a triple spectrometer (Spex Triplemate) comprising a double monochromator coupled for subtractive dispersion acting as a bandpass filter and a final spectrograph stage coupled to an intensified diode array (PARC 1420/1218) operated in conjunction with an OMA-2 console (PARC 1215) for storage and analysis of data. The intensifier of the diode array is gated to be turned on for 80 ns using a high voltage pulser (PARC 1211), during which the probe laser pulse occurs, thereby discriminating against emission from the sample caused by the photolysis pulse. The timing of the lasers and HV gate is accomplished by the use of coupled pulse and delay generators, and the system operated at a repetition rate of 10 Hz. Typically, one spectrum comprises 2400 pump-probe pulse pairs with the corresponding dark count background subtracted. The spectral and temporal accuracies of the system are ±6 cm^{-1} and ±10 ns respectively.
Ground state Raman spectra of the parent molecules were taken using a conventional scanning spectrometer with PM tube detection, and the 514.5 nm line of a cw Ar^+ laser. The spectral accuracy was ±2 cm^{-1}.

The sodium salts of AQ2S and AQ26DS (Aldrich) were purified by three-fold recrystallisation from water in the dark. The synthesis of AQ2MS has been described,[26] and it was used as received from Professor H.D. Scharf. The anthraquinone derivatives were used at a concentration of 5×10^{-3} M, and where present, sodium nitrite at a concentration of 0.1 M. Sodium hydroxide and sulphuric acid were used to achieve pH adjustments, and all solutions were oxygen saturated to avoid any steady state build up of intermediates.

3. RESULTS AND DISCUSSION

Observations in the presence of sodium nitrite. Of the two schemes for aqueous solution photochemistry detailed in the introduction, that in the presence of high concentrations of a strong electron donor (Figure 1) is the best characterised. The absorption spectra of $AQ^{\bar{\cdot}}$ and $AQH^{\cdot}$ are obtainable from samples in deoxygenated solutions using a conventional microsecond flash photolysis system, and their spectra have been well documented.[19-22] Both species exhibit weak absorption at wavelengths longer than 600 nm, but the main features ($\varepsilon_{MAX} > 6000\ M^{-1}\ cm^{-1}$) lie at ∿500 nm and ∿390 nm for $AQ^{\bar{\cdot}}$ and ∿400 nm for $AQH^{\cdot}$. Since these two semiquinone species are the most easily studied, initial experiments have centered on observation of their TR^3 spectra.

To ensure a high yield of radical anion and to avoid substantial production of transients B and C, a high concentration of a strong electron donor is required. In these experiments, sodium nitrite was used since its strong reducing nature is known from transient absorption measurements[23,24] to allow electron transfer to the triplet.

Nitrite ion was used at a concentration of 0.1 M, thereby ensuring a rate constant for the electron transfer two orders of magnitude greater than that for reaction with water.

Initially, it is important to establish that no photolysis is caused by the probe laser alone. This is confirmed by observation of the ground state spectra of the molecules using the probe laser under the conditions in which the TR^3 experiments were performed. Table 1 lists the AQ26DS bands observed using both cw and pulsed laser excitation. Only the strong bands observed using the conventional cw apparatus are seen in the spectrum obtained using the experimental arrangements for the TR^3 experiments. In addition to the bands listed in Table 1, a band at 1340 cm^{-1} is also observed when sodium nitrite is present, and this is assigned to the nitrite ion itself.

TABLE I Raman bands (wavenumber/cm^{-1}) of ground state AQ26DS obtained with cw and pulsed laser excitation.

514.5 nm cw		480 nm pulsed
1669	(vs)	1672
1596	(s)	1601
1481		
1420		
1323		
1281		
1177	(vs)	1182
1121		
1096		
1038		
957		

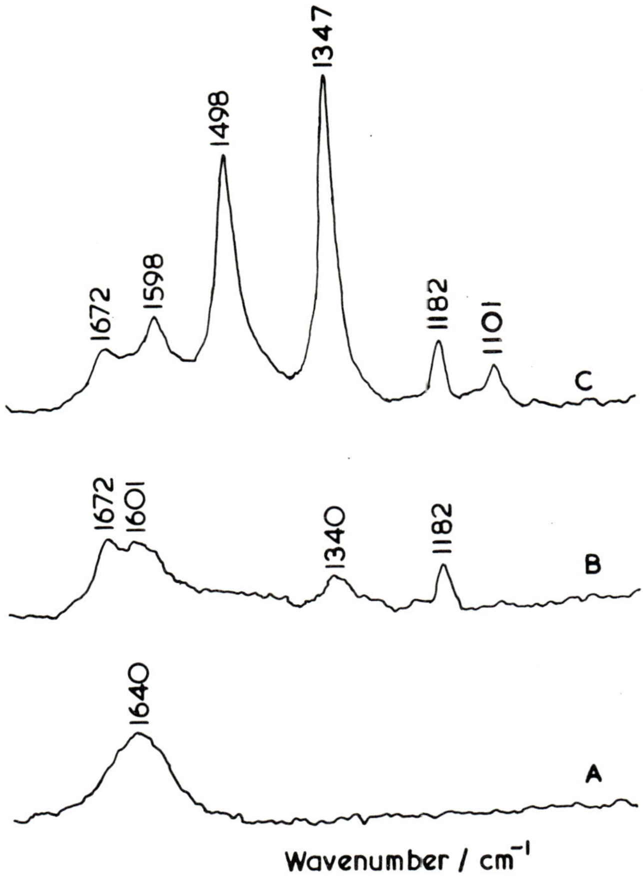

FIGURE 6 Raman spectra obtained using a probe laser wavelength of 480 nm of (A) water only, (B) AQ26DS and $NaNO_2$, and (C) TR^3 spectrum observed at a time delay of 100 ns after excitation at 337 nm.

Figure 6B shows the ground state spectrum of AQ26DS in the presence of sodium nitrite obtained using the pulsed laser at 480 nm, and Figure 6A the solvent (water) spectrum recorded under the same conditions. The TR^3 spectrum obtained at a time delay of 100 ns following 337 nm excitation is given in Figure 6C. The time

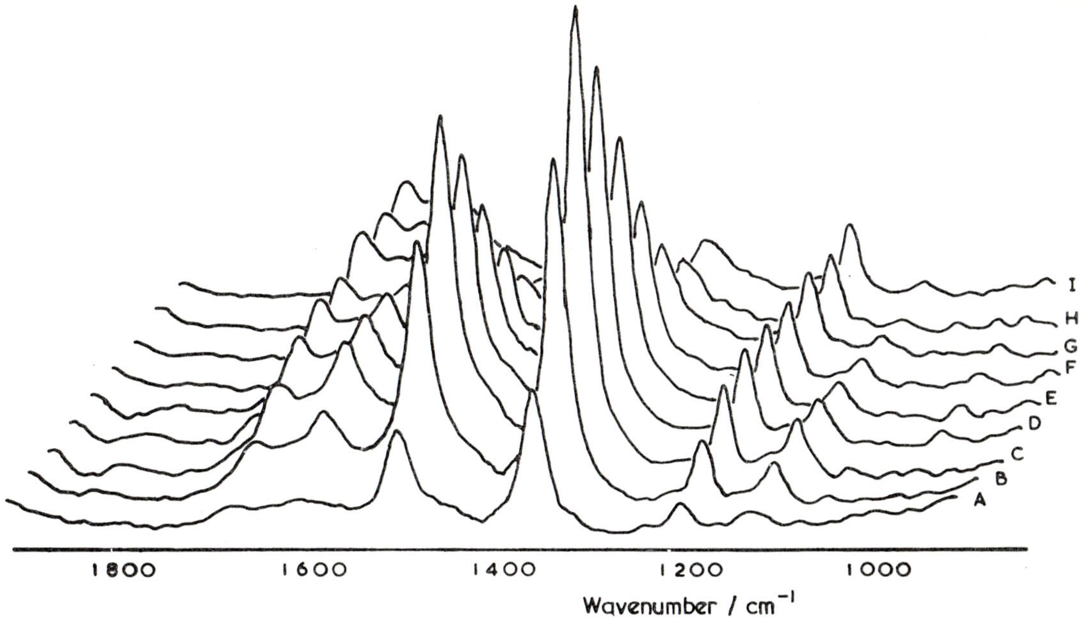

FIGURE 7 Time dependence of the transient spectrum of AQ26DS in the presence of $NaNO_2$ obtained using a pump wavelength of 337 nm and a probe wavelength of 480 nm. Time delays of: (A) 20 ns, (B) 50 ns, (C) 100 ns, (D) 500 ns, (E) 1 μs, (F) 2 μs, (G) 5 μs, (H) 10 μs, and (I) spectrum obtained with the probe laser only.

dependence of this spectrum can be studied by varying the time delay between the pump and probe laser pulses, and Figure 7 illustrates the changing intensities of the transient bands at various time delays. The spectrum recorded at long time delays is observed to be the same as that obtained using the dye laser alone, confirming that no signifi8ant permanent photolysis occurs. All four new bands observed in the TR^3 spectrum show the same time dependence, and therefore are assigned to the same transient species. Analysis of the kinetics of these four bands has given a pseudo-first-order decay rate constant of $5.1 \times 10^5\ s^{-1}$, which was used to calculate the second order rate constant for quenching by oxygen (1):

$$AQ26DS^{\overline{\cdot}} + O_2 \rightarrow AQ26DS + O_2^{\overline{\cdot}} \quad (1)$$

as $4.1 \times 10^8\ M^{-1}\ s^{-1}$. This value is in good agreement with values given in the literature for the quenching of the radical anions of sulphonated anthraquinones by oxygen.[27,28] Hence, the TR^3 spectrum in Figure 6C is assigned to $AQ26DS^{\overline{\cdot}}$.[29] Similar spectra were observed for $AQ2S^{\overline{\cdot}}$ and $AQ2MS^{\overline{\cdot}}$.

The pK_A of the system (2) is reported:

$$AQ26DS^{\overline{\cdot}} + H^+ \rightleftarrows AQ26DSH^{\cdot} \quad (2)$$

to be 3.2[18], and by adjusting the pH of the solution used to obtain the radical anion spectra to pH < 3, the neutral semiquinone radical should be obtained in appreciable yield. Using the probe laser at a wavelength of 480 nm results in no observable TR^3 spectrum at pH < 3 since under these conditions there is no species with a transient absorption spectrum at this wavelength

TABLE II TR^3 bands (wavenumber/cm^{-1}) assigned to the radical anions of the three molecules studied. *Raman bands also observed in the ground state spectra of the parent molecules.

AQ26DS$^{\overline{\cdot}}$	AQ2S$^{\overline{\cdot}}$	AQ2MS$^{\overline{\cdot}}$
1672*	1666*	1664*
1598	1597	1594
1498	1496	1498
1347	1344	1342
		1212*
1182*	1180*	1180*
1101	1085	1042

through which resonance enhancement of the Raman spectrum could occur. The strong absorption band of AQ26DSH$^{\cdot}$ centred at $\sim$ 400 nm does however allow new resonance Raman bands to be detected by tuning the dye laser to higher energies. Unfortunately, at wavelengths shorter than 440 nm two-photon effects occur since one dye laser pulse alone is observed to both pump the ground state sample into the excited state and probe the Raman spectrum of the resulting species. This occurs through excitation into the $S_1(n\ \pi^*)$ state, which is only weakly absorbing (ε < 100 M^{-1} cm^{-1}). This effect is seen through by the observation of Raman spectra, using the probe laser alone, which are not assignable to groundstate AQ26DS spectra observed using the cw system or the dye laser at longer wavelengths (see Figure 6B and Table 1). For wavelengths longer than 420 nm, this problem can be overcome by attenuating and defocussing the lasers thereby reducing the power density at the sample, and hence experiments have been limited to probe laser wavelengths longer than 420 nm and power densities which do not result

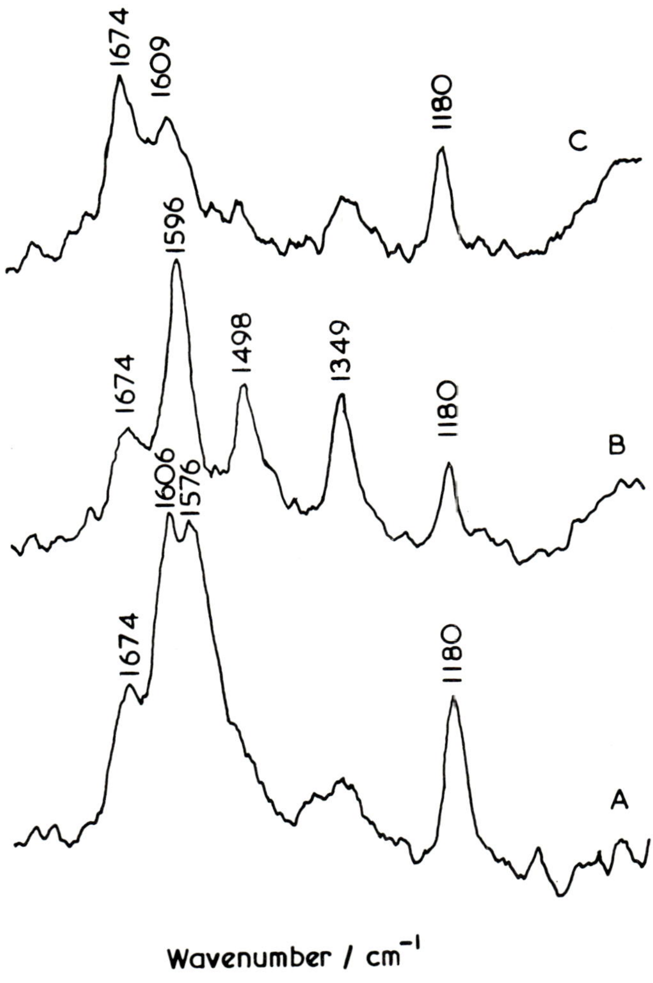

FIGURE 8 Raman spectra of AQ26DS in the presence of $NaNO_2$ obtained using a pump wavelength of 351 nm and a probe wavelength of 430 nm. (A) TR^3 spectrum at pH1 and a time delay of 1 μs between the pump and probe pulses, (B) TR^3 spectrum at pH6 and a time delay of 1 μs, and (C) spectrum obtained with the probe laser only.

in the observation of two-photon effects. Consequently, although the main feature of the neutral radical absorption spectrum is centred at ∿ 400 nm, attempts to observe the TR^3 spectrum have been limited to probe laser wavelengths to the red of this band. Figure 8 illustrates spectra taken at a probe wavelength of 430 nm of AQ26DS in the presence of sodium nitrite. Figure 8C shows the ground state spectrum obtained with the dye laser only, which is observable between pH1 and pH7, and which is comparable with ground state spectra obtained under cw excitation. Figures 8A and 8B show the TR^3 spectra obtained 1 μs after excitation at 351 nm at pH1 and pH6 respectively. The spectrum observed at pH6 is that of $AQ26DS^{\bar{\cdot}}$, although there is a large difference in the relative band intensities compared with observations at 480 nm because at this probe laser wavelength the enhancement is gained from resonance with the absorption band centred at ∿ 390 nm, i.e. a different electronic state. The 1101 cm^{-1} band is not observed, but the other TR^3 bands remain the same within experimental error. At pH < 3, the $AQ26DS^{\bar{\cdot}}$ bands reduce in intensity and at pH1 are not detected. They are replaced by two bands at 1606 cm^{-1} and 1576 cm^{-1} which are assigned to $AQ26DSH^{\cdot}$.

4. OBSERVATIONS IN PURE AQUEOUS SOLUTION

The study of the sulphonated anthraquinones in pure aqueous solution is much more difficult because of the presence of several transients simultaneously (Figure 2), as was stated above. The triplet has been observed using various probe laser wavelengths in the range 420 nm to 480 nm, and identified by the presence of bands which decay on a relatively short timescale of less than 1 μs. For AQ26DS

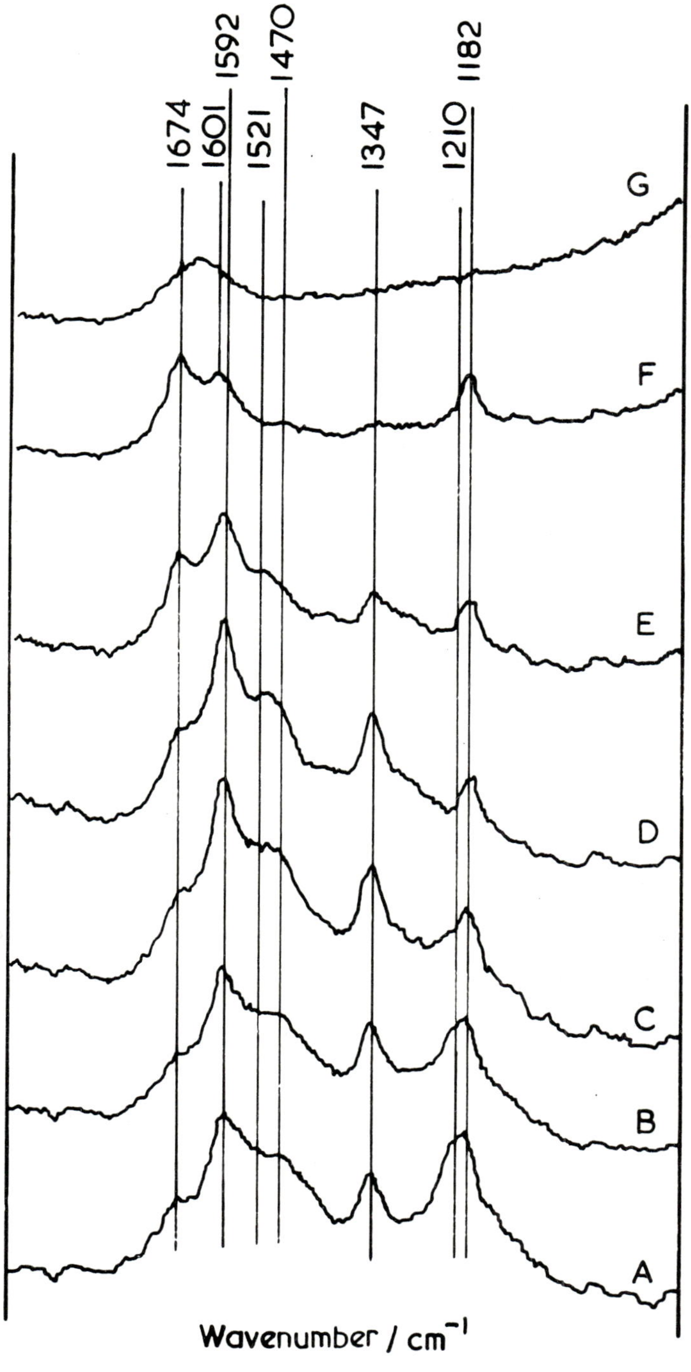

FIGURE 9 Raman spectra of AQ26DS in aqueous solution obtained using a pump wavelength of 337 nm and a probe wavelength of 441 nm. Time delays of: (A) 20 ns, (B) 75 ns, (C) 200 ns, (D) 1 μs, (E) 10 μs, and spectra obtained with the probe laser only of (F) AQ26DS in aqueous solution, and (G) water only.

the decay lifetime is approximately 500 ns, and is in good agreement with measurements of the triplet lifetime in aqueous solution made using time-resolved absorption and emission spectroscopy.[24] The triplet-water complex labelled transient B in the flash photolysis studies, and which was observed in those studies to grow in concurrently with decay of the triplet and then decay over several microseconds, has also been observed in the TR^3 experiments. Spectra of this species have been observed using probe laser wavelengths between 430 nm and 480 nm, and have been identified by the presence of bands which follow the same time dependence as the transient absorption spectrum of B, growing in intensity over approximately 500 ns and then decaying on the microsecond timescale. A series of spectra taken at different time delays between the pump and probe lasers, where the pump and probe wavelengths are 337 nm and 441 nm respectively, illustrates this point. The time dependence of the bands leads to the tentative assignments given in Table 3. Similar observations have also been made in studies on AQ2S and AQ2MS.

TABLE III Tentative assignment of the bands (wave (wavenumber/cm^{-1}) of AQ26DS observed in pure aqueous solution (Figure 9).

AQ26DS	^{3}AQ26DS	F(AQ26DS)
1674	-	-
1601	-	-
-	-	1592
-	-	1521
-	1470	-
-	-	1347
-	1210	-
1182	-	-

As stated above, two photon effects are observed in the ground state spectra of the parent molecules at probe laser wavelengths where the photons can be absorbed by the molecules. A similar effect has been observed with the spectra of transient B. The TR^3 spectra produced under the same conditions used to obtain those illustrated in Figure 9, but with the probe laser tuned to the absorption maximum of transient B at 480 nm, are found to be identical to the radical anion spectra observed on electron transfer from nitrite ion, but with the same time dependence as transient B. By attenuating and defocussing the lasers and reducing the power density at the sample, however, the spectrum changes and at low power densities the 1498 cm^{-1} peak of $AQ26DS^{\overline{\cdot}}$ is replaced by peaks at 1550 cm^{-1} and 1521 cm^{-1} whilst the 1592 cm^{-1}, 1347 cm^{-1}, and 1180 cm^{-1} peaks remain in the same position. The spectrum in Figure 10A is assigned to transient B and displays the same time dependence as the spectrum observed at a probe laser wavelength of 441 nm. That the radical anion spectrum observed at high power densities has the same lifetime as transient B indicates that $AQ26DS^{\overline{\cdot}}$ is not produced by the pump pulse since in this case it would display the shorter lifetime of $AQ26DS^{\overline{\cdot}}$ produced by electron transfer. Since it displays the same lifetime as B, it appears that B is produced as expected and has the predicted time dependence, but that the probe laser pulse causes photolysis to produce the radical anion and probing of its spectrum within the duration of a single pulse. Preliminary attempts to observe TR^3 bands of transient C at longer probe laser wavelengths have indicated that a similar effect is seen with this species and that $AQ26DS^{\overline{\cdot}}$ spectra are observed at high power

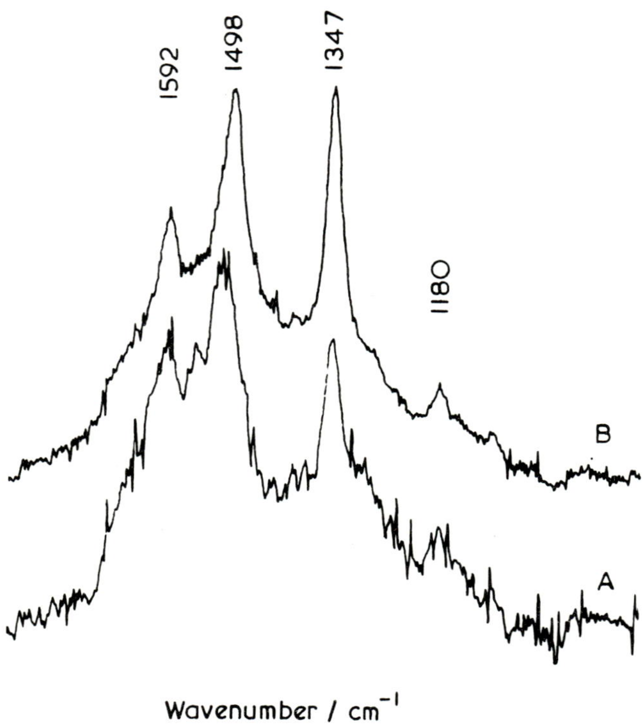

FIGURE 10 TR^3 spectra of AQ26DS in aqueous solution obtained using a pump wavelength of 351 nm, a probe wavelength of 480 nm, and a time delay of 1 μs between the two pulses. (A) lasers focussed closely below the sample, and (B) lasers defocussed, scale x 3.

densities. Again, similar spectra are observed for AQ2S and AQ2MS. The observation of the radical anion spectrum on photolysis of B and C indicates the charge-transfer nature of these transients, with the strongly oxidising nature of the triplet attracting electron density from water to form the complex and absorption of a photon by the complex leading to complete electron transfer and production of the radical anion. It has been suggested that the two complexes are the carbonyl and benzenoid ring adducts of water to the triplet,[23] and it is hoped that analysis of the TR^3 spectra should lead to the confirmation or refutation of this suggestion.

5. CONCLUSION

TR^3 spectroscopy has been used to confirm the presence of several transient species produced following photolysis of sulphonated anthraquinones in oxygenated aqueous solution. The transient Raman spectra differ significantly from the ground state spectra of the parent molecules, and the kinetic properties of the TR^3 bands have been used to assign them to species initially observed by their optical absorption properties. Although the three 'strong' sensitizers studied have widely different kinetic properties, and for AQ2MS different transient absorption spectra also, it appears from initial studies that the Raman spectra of their transients are similar. Further work is in progress to obtain data of improved quality and also to study the polarization properties of the bands observed. It is hoped that this will aid in the assignment of the bands to specific vibrations and thereby in the elucidation of the structures of the transients.

6. ACKNOWLEDGEMENTS

These experiments were performed at the UV Radiation Facility of the Rutherford Appleton Laboratory, and we thank Dr. Fergus O'Neill, Mrs. Jan Szechi and Mr. Graham Hogg for their assistance. We thank Mr. R.B. Girling for obtaining the cw Raman spectra, Professor H.D. Scharf for his gift of AQ2MS, and S.E.R.C. for financial support, in particular the award of a Senior Visiting Fellowship to GHA.

REFERENCES

1. J.M. BRUCE, in The Chemistry of the Quinonoid Compounds, edited by S. Patai (Wiley, New York, 1974).
2. A.V. EL'TSOV, O.P. STUDZINSKII and V.M. GOEBENKINA, Russ. Chem. Rev., (English trans.), 46, 93 (1977).
3. J.L. BOLLAND and H.R. COOPER, Proc. Roy. Soc. A, 225, 405 (1954).
4. J.J. MORAN and H.I. STONEHILL, J. Chem. Soc, 765, 779 and 780 (1957).
5. N.K. BRIDGE, J. Soc. Dyers Col., 76, 484 (1960).
6. G.O. PHILLIPS and J.C. ARTHUR, Textile Res. J., 34, 497 and 572 (1964).
7. J.F. McKELLAR, Radiat. Res. Rev., 3, 141 (1971).
8. J.F. McKELLAR and G.O. PHILLIPS, Ind. Eng. Chem. Prod. Res. Dev., 19, 23 (1980).
9. H.D. SCHARF and R. WEITZ, Tetrahedron, 35, 2255 (1979).
10. I. OKURA and N. KIM-THUAN, Chem. Lett., 1569 (1980).
11. B. MOONEY and H.I. STONEHILL, Chem. and Ind., 1309 (1961).
12. K.P. CLARK and H.I. STONEHILL, J. Chem. Soc. Faraday Trans. 1, 68, 577 (1972).
13. K.P. CLARK and H.I. STONEHILL, J. Chem. Soc. Faraday Trans. 1, 68, 1676 (1972).
14. A.D. BROADBENT and R.P. NEWTON, Can. J. Chem., 50, 381 (1972).
15. A.D. BROADBENT, H.B. MATHESON and R.P. NEWTON, Can. J. Chem., 53, 826 (1975).
16. K. TICKLE and F. WILKINSON, Trans. Faraday Soc., 61, 1981 (1965).
17. A.A. LAMOLA and G.S. HAMMOND, J. Chem. Phys., 43, 2129 (1965).
18. A. HARRIMAN and A. MILLS, Photochem. Photobiol., 33, 619 (1981).
19. C.F. WELLS, Trans. Faraday Soc., 57, 1703 and 1719 (1961).
20. N.K. BRIDGE and G. PORTER, Proc. Roy. Soc. A., 244, 259 (1958).
21. V.A. KUZMIN and A.K. CHIBISOV, Chem. Comm., 1559 (1971).
22. J. METCALFE, J. Chem. Soc. Faraday Trans. 1, 79, 1721 (1983).
23. I. LOEFF, A. TREININ and H. LINSCHITZ, J. Phys. Chem., 87, 2536 (1983).

24. J.N. MOORE, D. PHILLIPS, N. NAKASHIMA and K. YOSHIHARA, (to be submitted for publication).
25. G.H. ATKINSON, in Advances in Infrared and Raman Spectroscopy, edited by R.J.H. Clark and R.E. Hester (Heyden, London, 1982), Vol. 9, p. 1.
26. H.D. SCHARF and R. WEITZ, Tetrahedron, 35, 2267 (1979).
27. B.E. HULME, E.J. LAND and G.O. PHILLIPS, J. Chem. Soc. Faraday Trans. 1, 68, 1992 (1972).
28. K.B. PATEL and R.L. WILLSON, J. Chem. Soc. Faraday Trans. 1, 69, 814 (1973).
29. J.N. MOORE, G.H. ATKINSON, D. PHILLIPS, P.M. KILLOUGH and R.E. HESTER, Chem. Phys. Lett., 107, 381 (1984).

CHAPTER 6

A TIME-RESOLVED RESONANCE RAMAN INVESTIGATION OF THE EFFECT OF TRYPTOPHAN-MODIFICATION ON PHOTOCYCLING EFFICIENCY AND CHROMOPHORE CONFORMATION IN BACTERIORHODOPSIN

P. J. ROGERS, A. E. W. KNIGHT and J. P. HARDY
Department of Chemistry,
Griffiths University,
Brisbane,
Australia.

1. INTRODUCTION

Time-resolved resonance Raman spectroscopy (TR^3) has proven to be a remarkable tool in elucidating the changing vibrational structure of the retinyl chromophore of the light-sensitive membrane protein, bacteriorhodopsin (bR). The protein acts as a light-driven trans-membrane proton pump in the photosynthetic bacterium, _Halobacteria halobium_. Extensive Raman spectroscopic studies, along with fluorescence and absorption spectroscopy of retinal, its Schiff base and its protein complexes have been carried out (for a review, see references 1 through 5) in an effort to understand how the protein influences retinal spectroscopy, and how retinal, in turn, induces the protein conformational changes required to initiate the proton pump function.

bR may be described as an 'isomerase'; that is, the interaction of the protein with retinal significantly lowers the potential barrier to trans-cis isomerization about the C-13 double bond of the polyene chain backbone of retinal, to the extent that, in the dark, a thermal equilibrium can be set up consisting of 50% of the all-trans conformer and 50% of the 13-cis conformer. Free all-trans retinal is essentially stable in the dark,[6] and light induced isomerization of retinal bound to bR proceeds with a quantum efficiency of 0.33,[7] an order of magnitude greater than the quantum efficiency for free retinal.

The key to understanding the mechanism by which the bR proton pump is activated clearly lies in an analysis of the protein-induced perturbations to the retinyl chromophore's ground and excited state structures which so enhance the quantum efficiency of isomerization. We decided to investigate the possibility of the involvement of specific protein-chromophore interactions in causing facile isomerization by chemical modification of the protein, namely oxidation of the indole moiety of tryptophan residues by n-bromo-succinimide (NBS). Tryptophan has been implicated as a potential mediator of retinal conformation in theoretical modelling of the binding site,[8] in experimental evidence that photocycling is impeded by tryptophan oxidation[9] and by the observation that aromatic residues in the bR sequence are perturbed during the bR photocycle.[10,11,12] Point charge models of the binding site[8,13] suggest that a dipolar, polarizable amino acid such as tryptophan is positioned with its negative electron density near the ionone ring of retinal. The absorption of light by the chromophore induces a shift of electron density away from the ionine ring towards the Schiff base linkage.[14] Hence, a suitably placed negative charge would stabilize the positive charge in the ring.

The changing interplay between the retinyl chromophore and its protein matrix following the absorption of light is evidenced by the photocyclic behaviour of bR shown in Figure 1. The symbols representing each transient intermediate of the photocycle includes the absorption maximum in nanometers which characterizes that intermediate. The normal photocycle which activates the proton pump is indicated by the trans cycle, so called to indicate that retinal isomerizes from its all-trans conformation in the light-adapted rest state (bR^{LA}_{570}) to 13-cis retinal in M_{412}. The large blue shift from 570 nm to 412 nm is caused mainly by deprotonation of the nitrogen of the Schiff base bond which covalently links retinal to the protein. This

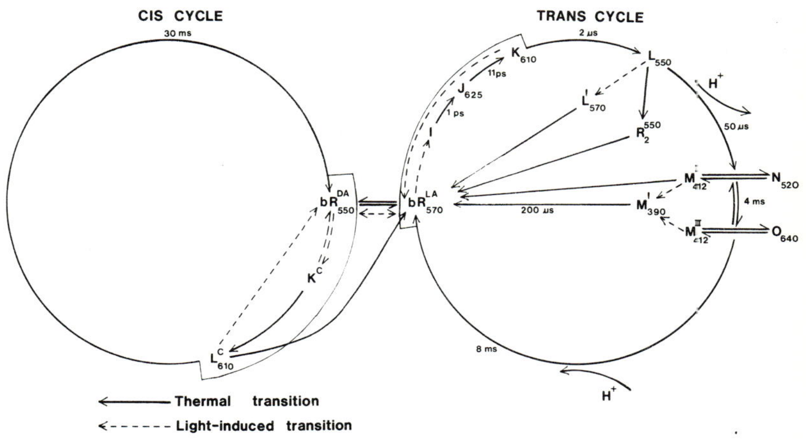

FIGURE 1 Diagrammatic representation of the normal 'trans' cycle and the 'cis' cycle of bR. The diagram utilizes the findings reported in references 2, 6, 15, 16 and 17.

suggests that the Schiff base nitrogen may be participating directly in the proton pump. In the dark, a thermal equilibrium between light-adapted (all-trans) and dark-adapted (13-cis) bR exists. Absorption of light by the dark-adapted form, bR_{550}^{DA}, activates the 13-cis cycle for which the Schiff base is not deprotonated and proton pumping does not occur. With continued irradiation, however, the cis cycle eventually returns all bR to its optimum rest state, bR_{570}^{LA}.

Perturbations to the photocycle, resulting in various branching reactions which short circuit the normal photocycle (i.e. bR_{570}-->K-->L-->M-->bR_{570}) are induced by extreme environmental conditions[2] of pH, temperature or lipid environment, or by high light intensities where a bR molecule absorbs more than one photon during the lifetime of the cycle (∿10 ms). Of particular relevance to the experimental techniques used for TR^3 to be reported here is the observation that a photostationary equilibrium can be created between bR_{570}^{LA} and K_{610} at sufficiently high light intensities.[18] Another consequence of multiple photon absorption during the lifetime of K_{610} (∿2 μs) is the light-induced coupling of the cis to the trans cycle.[15,16] It should be noted that the quantum yield of this coupling is normally very low and can only be observed in native bR when it is pumped by relatively high intensity radiation from lasers or high power Xe flash lamps.

The yield of M_{412} is frequently used as a measure of photocycling efficiency since the deprotonation of the Schiff base linkage is implicated in the proton pump mechanism. Figure 2 shows the effect on M_{412} yield when tryptophan residues are successively oxidized by

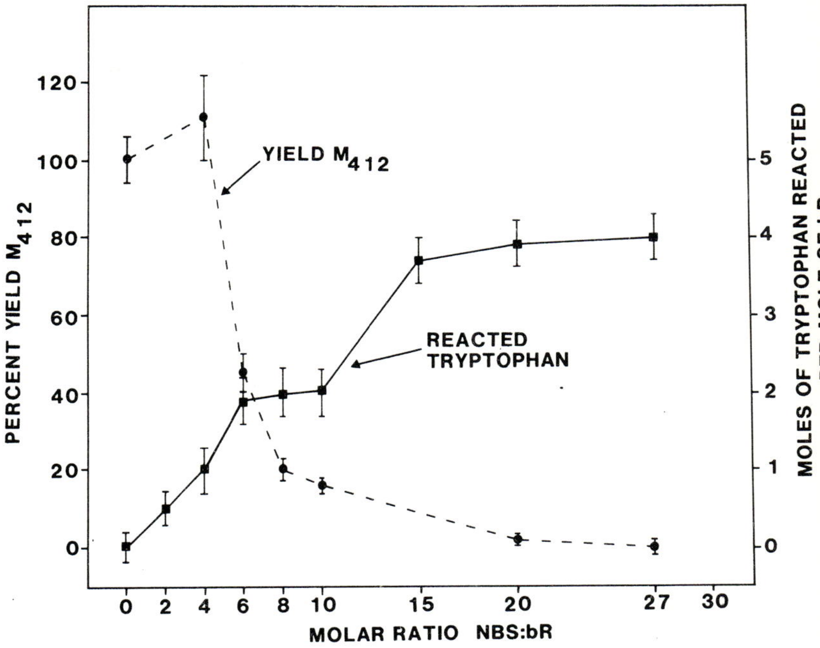

FIGURE 2 Photocycling yield of M_{412} for NBS-bR for various molar reacting ratios NBS:bR. M_{412} yield was measured from the maximum change in absorbance of a continuous probe beam (410±10 nm) following an actinic flash from a flashlamp-pumped dye laser (400 ns, 100 mJ, 570 nm). Protein concentration was 20 μM in distilled H_2O, pH 6. Also shown is the amino acid analysis of tryptophan content of the samples used for M_{412} yield measurements.

increasing the NBS:bR molar ratio. Also shown in this figure is the result of amino acid analyses of the samples used in the flash photolysis experiment to determine photocycling efficiency. With one mole of tryptophan oxidized per mole of bR, there is a small increase in M_{412} yield relative to an unmodified control sample. A similar rise in yield was observed by Konishi and Packer in their initial study of NBS-modified bR (NBS-bR).[9] However, when a second mole of tryptophan per mole of bR is oxidized (NBS:bR 8:1 to 10:1) the yield drops to less than one third that of the control.

Amino acid analyses of modified and native bR in membrane fragments were carried out to locate the sites susceptible to NBS reaction. Figure 3 summarizes the fragmentation process used in identifying these sites. A more detailed description of experimental techniques and results of amino acid analyses of bR is presented in reference 9. As will be seen from examination of Figure 3, the second tryptophan accessed may be any one of six found on the C_1 fragment (residues 72 to 248) produced by cleavage with chymotrypsin. Only tryptophans 10 and 12 on the C_2 fragment (residues 1 to 71) are reacted for NBS:bR 5:1 and Figure 2 shows that the oxidation of one of these residues appears to have no adverse effect on photocycling efficiency. Of the six remaining tryptophan residues in bR, TRP-80 or TRP-86 (or perhaps both) shows the greatest susceptibility to reaction with NBS. However, small losses of TRP-137, 138, 182 and 189 cannot be discounted. This means that the decreased photocycling efficiency induced by oxidation of a second tryptophan in the bR sequence appears to be due to the cumulative effects of tryptophan modification, possibly at several sites in the 3-D structure of bR, rather than to the modification of a specific tryptophan involved in a particular function in the mechanism of proton translocation, either as a proton acceptor or as a mediator of retinal conformation.

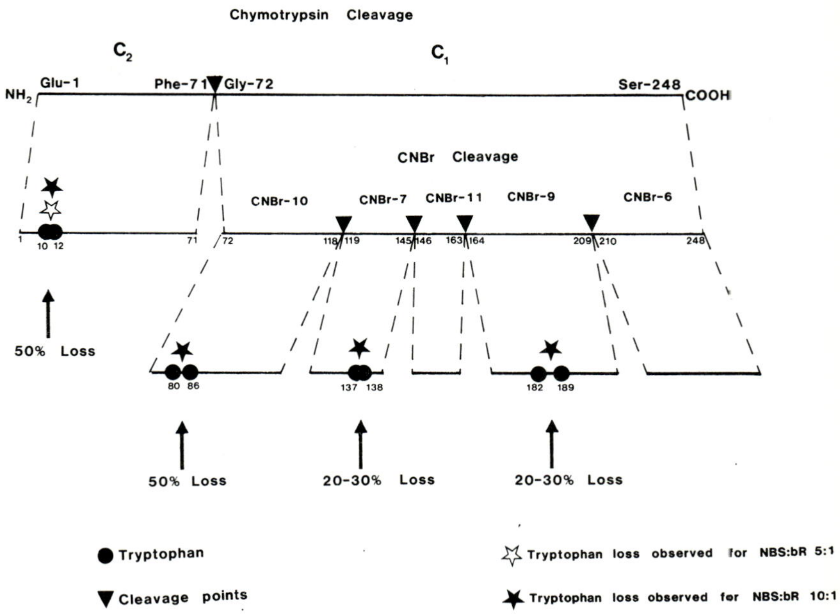

FIGURE 3 Summary of the fragmentation process used to locate modified tryptophan residues. Chymotrypsin cleaves the bR peptide chain between PHE-71 and GLY-72 resulting in two fragments, C_1 (residues 72 to 248) and C_2 (residues 1 to 71). Further cleavage of C_1 at methionine residues with CNBr produces 5 fragments, three of which contain two tryptophan residues each.

It should be noted at this stage that any study such as this, in which the functional importance of particular amino acid residues in a protein is investigated by chemical modification, must take into consideration other effects of the reaction which may include gross perturbations to tertiary as well as secondary protein structure.

In this study, where retinal confirmation following tryptophan modification was to be probed by TR^3 and other spectroscopic techniques, the nature of the perturbation to the system had to be relatively small - that is, the changes in efficiency should ideally be traceable to definable changes in protein structure and/or internal residue interactions. Gross changes to tertiary structure will so disrupt overall protein morphology that normal functioning is completely halted and the modified protein resembles very little the native form. Accompanying the TR^3 study of modified membranes to be described below was an extensive investigation of the structural implications of NBS modification of bR. A complete description of this work is beyond the scope of this paper, and is reported elsewhere.[19] A brief summary of results and conclusions from this work is presented below to conclude this introductory section.

Diagrammatic representations of bR structure as based upon the findings of Engelman and coworkers,[20] and Argard and Stroud,[21] are presented in Figures 4 and 5. Figure 4 is a plan view of the 7 α-helices that comprise the bulk of the bR secondary structure. The helices are arranged almost perpendicularly to the membrane with retinal attached to lysine-216 on the helix containing the carboxyl terminus. Two probable configurations are shown; recent evidence[21] suggests that Model I is the most probable. The ionone ring of retinal appears to be held in the plane of the membrane as shown.[22] Three bR molecules group around the C_3 axis to form extended hexagonal arrays in the membrane.

Figure 5 shows the approximate transverse locations of tryptophan residues in the plane of the membrane. All

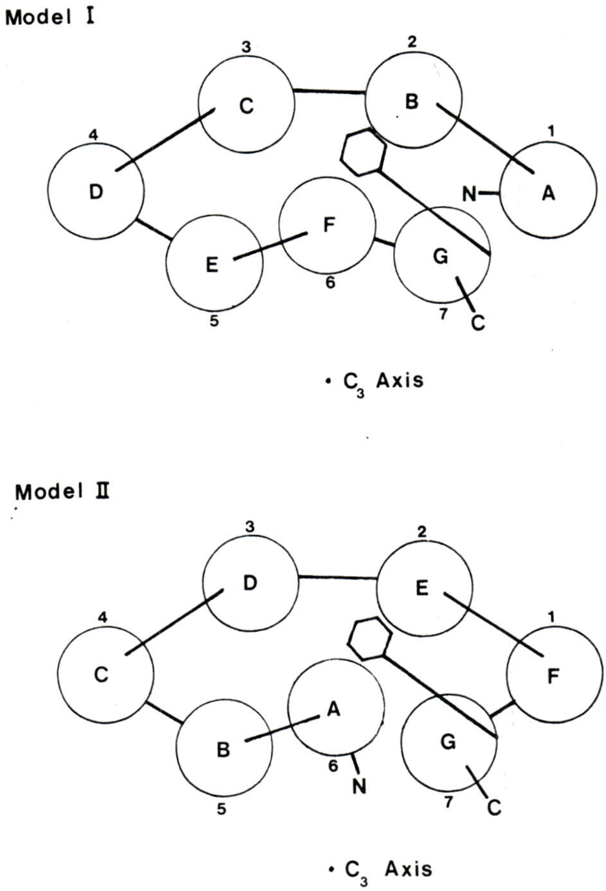

FIGURE 4 Plan view of the two arrangements of the 7 α-helices of bR which satisfy experimental evidence[20,21] regarding linking regions between the α-helices and the location of the carboxyl terminus. Retinal is located according to the results outlined in reference 22. The position of the C_3 axis is also shown. Model I is favoured over model II because this arrangement of the α-helices places internal charged residues in closer contact.

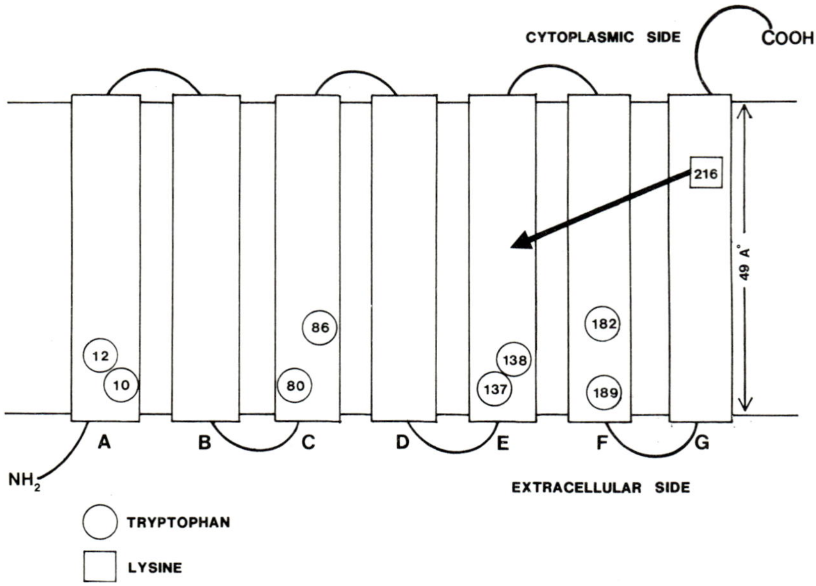

FIGURE 5 The approximate locations of tryptophan residues in the α-helices of bR. The location of LYS-216 which binds to retinal to form the purple complex is also shown. The arrow indicates the approximate orientation of retinal with respect to the plane of the membrane. Probable arrangements of helices A to G are shown in Figure 4.

are located close to the external surface and close to linking regions of the protein which connect the α-helices together. Retinal, indicated by the arrow, is inclined at an angle of about 23° [22] to the membrane plane with the ionone ring towards the external surface and perhaps in close contact with some of the tryptophan residues, particularly those of helix C.

With this information about the structure of bR, some interpretation is possible of the CD fluorescence and absorption spectra with regard to structural changes induced by NBS. The far UV CD spectrum of modified membranes indicates that localized unfolding of the predominantly α-helical secondary structure occurs. Accompanying this unfolding appears to be some relaxation of the tertiary structure. This conclusion was drawn from the observation of increased polarity of the environment of some internal aromatic residues as indicated by changes in the near UV absorption and CD spectra.[19] This is consistent with some of the α-helix columns comprising the tertiary structure changing their relative orientations or proximity to one another, and thereby allowing greater penetration of H_2O to the mainly hydrophobic protein interior.

The induced optical activity of retinal bound to the active site of the protein is also reduced when tryptophan residues are oxidized. Decreased rotational strengths in the near UV and visible regions indicated reduced protein-retinal interaction and protein-protein interaction in the normal hexagonal trimeric array. This conclusion is supported by the absence in near UV absorption and fluorescence spectra of any evidence for the existence of planar ring-chain conformations for retinal in NBS:bR 5:1 (1 tryptophan modified) both show evidence for the retinyl chromophore being rigidly constrained in a planar ring-chain configuration. The chromophore of NBS:bR 10:1 has spectral characteristics closer to those displayed by free solution retinyl moieties for which limited rotation about the single bond connecting the ionone ring to the polyene chain occurs. The combined evidence from flash photolysis and CD fluorescence and absorption spectroscopy of NBS:bR suggests that the reduction in yield accompanying the modification of two tryptophans on average in the bR sequence can be linked with increased retinal mobility in its protein matrix. In other words, the protein constraints upon retinal conformation, which are postulated to account for the spectroscopic properties of bound retinal as compared to free Schiff bases of retinal, have been reduced by tryptophan modification.

Raman spectroscopy can provide further information about NBS:bR in two ways: firstly, the resonance Raman process allows a more detailed determination of changes to retinal conformation with NBS modification, and secondly, the introduction of time resolution allows the further investigation of changes to the mechanism of photocycling by observation of different intermediates in the cycle. The aim of the TR^3 investigation of NBS:bR compared to unmodified bR was to establish whether a link exists between photocycling efficiency and chromophore structure as mediated by protein interactions.

1.1. CW Raman Spectroscopy

1.1.1. Results and observations. As a precursor to the main time-resolved work using pulsed excitation, flow resonance Raman spectra of NBS-bR were recorded using Ar^+ ion laser excitation at 514.5 nm, 488.0 nm and 476.5 nm. Results are shown in Figures 6, 7 and 8. In these figures, the steady state flow Raman spectra of NBS:bR 8:1 or 10:1 are compared with native bR spectra recorded for the same conditions of protein concentration, sample flow rate and incident power.

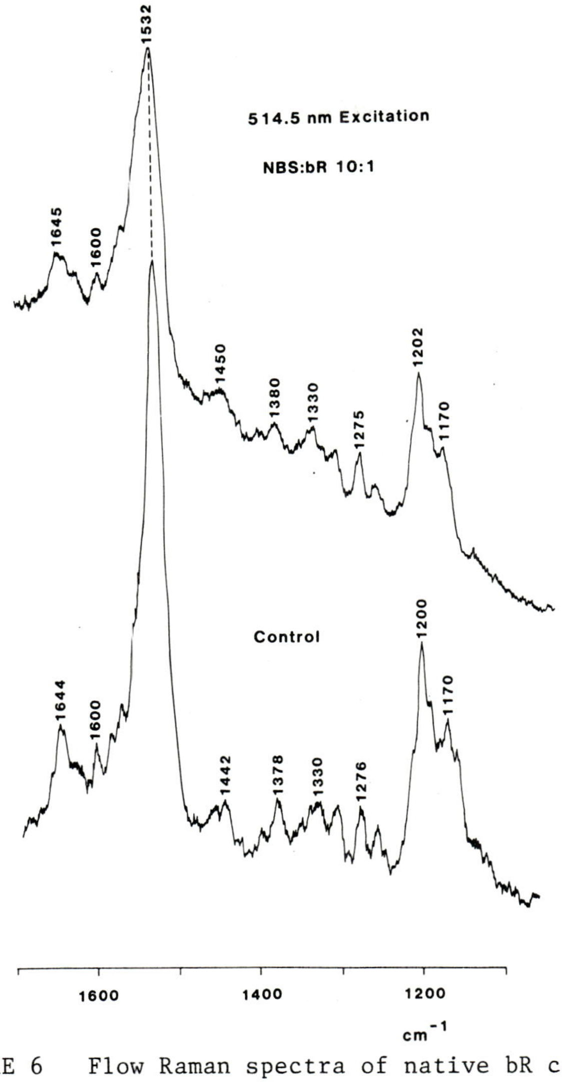

FIGURE 6 Flow Raman spectra of native bR control and NBS:bR 10:1 using Ar^+ ion CW excitation at 514.5 nm. Spectral resolution, 3 cm^{-1}; incident power, 1 W; protein concentration, 20 μM. Samples were flowed continuously from a 7 ml reservoir to minimize photodegradation.

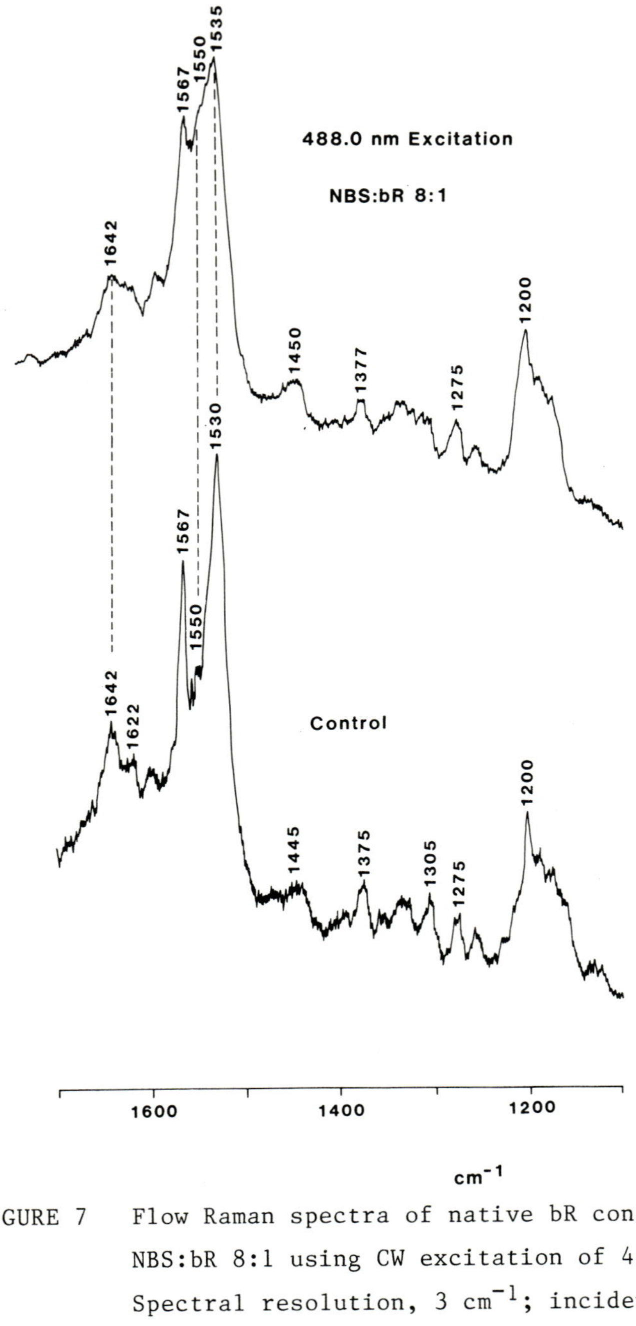

FIGURE 7 Flow Raman spectra of native bR control and NBS:bR 8:1 using CW excitation of 488.0 nm. Spectral resolution, 3 cm^{-1}; incident power, 1 W; protein concentration, 20 μM.

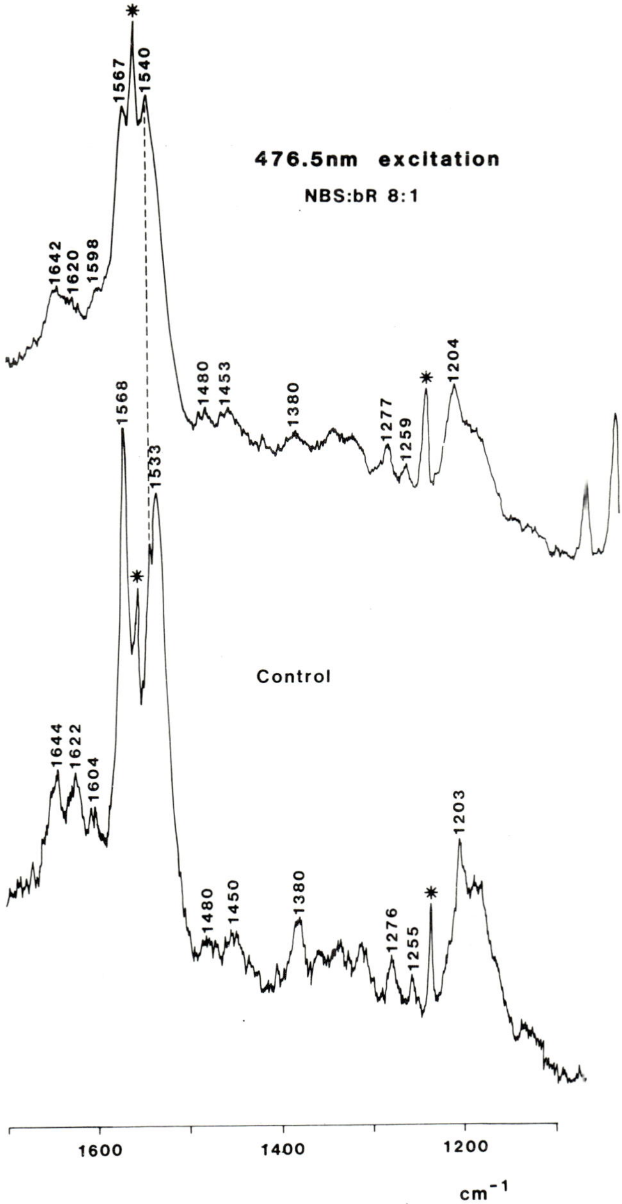

FIGURE 8 Flow Raman spectra of native bR control and NBS:bR 8:1 using CW excitation at 476.5 nm. Spectral resolution, 4 cm^{-1}; incident power, 200 mW; protein concentration, 20 μM.

The Raman spectrum of native bR obtained with 514.5 nm excitation (Figure 6) displays predominantly bR_{570} characteristics due to the resonance enhancement of bR_{570}, despite the fact that significant pumping of the photocycle occurs at this wavelength. bR_{570} is characterized in these spectra by an intense ethylenic stretch band at 1532 cm^{-1} (Table 1) and the protonated Schiff base vibration at 1644 cm^{-1}. The same region for NBS:bR 10:1 shows extra intensity appearing on the high energy side of the ethylenic peak. There is also a small loss of intensity at 1644 cm^{-1} and an accompanying increase, relative to native bR, in the weaker 1620 cm^{-1} region that corresponds to the signature for the deprotonated Schiff base. Changes occurring in the fingerprint region are small and are more apparent with excitation at 476.5 nm and 488.0 nm.

Scattering from M_{412} is strong with excitation at these shorter wavelengths (Figures 7 and 8) due to resonance enhancement of this intermediate over bR_{570}. A strong peak at 1567 cm^{-1} and the appearance of a definite band at 1622 cm^{-1} indicate a significant contribution to Raman intensity from M_{412}. For NBS-bR there is again new intensity appearing between 1530 cm^{-1} and 1567 cm^{-1} and loss of definition of the 1644 cm^{-1} and 1622 cm^{-1} bands. In the fingerprint region, the 1305 cm^{-1} band is likewise reduced in intensity and significantly broadened. The 1200 cm^{-1} peak also shows additional intensity to the high energy side.

These changes are most apparent in the case of excitation at 476.5 nm. The peaks marked with an asterisk (*) are Rayleigh scattering from laser plasma lines. For NBS:bR 8:1 the 1533 cm^{-1} peak has been completely dominated by a peak at 1540 cm^{-1} which normally appears as a shoulder to

TABLE 1 ν(-C = C-) frequencies (cm^{-1}) for intermediates of the bR photocycle

Intermediate	ν(-C = C-) (cm^{-1})	Reference
bR^{LA}_{570}	1530 - 1532	
K_{610}	1516	25
	1518	17
	1524	26
	1528	27
L_{550}	1537	17,35
	1547	36
X	1552	35
bR^{DA}_{550}	1530	35
	1536	36
	1537	17
M_{412}	1567	17

ν(-C = C-) Assignments for Branching Intermediates from Stockburger et al.[17]

R_1^{570} (= K^C)	1530	photoproduct of K_{610} in cis cycle
$R_1^{'590}$ (= L^C_{610})	1522	relaxation product of R_1^{570} in cis cycle
L'_{570}	1530	photoproduct of L_{550}
R_2^{550}	1537	thermal relaxation product of L_{550}, bypassing M_{412}
M'_{390}	∿1563?	photoproduct of M_{412}

the ethylenic stretching peak of bR_{570}. There is also loss of definition in the Schiff base vibration region and the fingerprint region, with particular loss of intensity in the 1380 cm^{-1} band. It is evident that there is less M_{412} produced since the 1567 cm^{-1} peak is reduced in intensity.

The changes described for these excitation wavelengths are only evident for NBS:bR 8:1 or 10:1 where two or more tryptophans have been modified in the bR sequence. NBS:bR 4:1, for which one tryptophan has been accessed, produces Raman spectra which are indistinguishable from those of native bR (results not shown).

In a CW excitation experiment such as is outlined here, where the pump and probe radiation is provided by the same beam, the spectrum recorded will be a function of the pump efficiency of the excitation wavelength as well as resonance enhancement effects. As Figure 1 shows, the bR photocycle is not a simple unidirectional cycle of intermediates. Thus, in the steady state CW Raman experiment, where multiple photon absorption is likely, the resultant band intensities will be a complex function of resonance enhancement of the intermediates present in the excitation volume as well as the relative concentrations produced by pumping of the photocycle and depopulation of certain states by further photolysis.

All spectra of NBS-bR obtained with a particular excitation wavelength, are compared with equivalent native bR spectra recorded under the same conditions of protein concentration and laser power. This means that differences between the spectra of native and modified bR reflect changes in steady state populations that are induced by tryptophan modification, and not by changes in incident light intensity, or sample flow rate. Spectra of native bR obtained using these flow techniques are in close agreement with those already reported using similar techniques.[23,24]

Excitation at 514.5 nm (Figure 6) resonance enhances the bR_{570} complex so that the Raman spectrum is dominated by the characteristic scattering of bR_{570}. The primary intermediate K_{610} has been assigned an ethylenic stretch frequency from 1516 cm^{-1} to 1528 cm^{-1} (see Table 1). Our native bR spectrum exhibits a strong ethylenic stretch peak at 1532 cm^{-1} which can be assigned to bR_{570}, with very little structure to the low vibrational energy side that could be associated with the primary intermediate. The peak is asymmetric with intensity appearing particularly in the 1540 cm^{-1} to 1550 cm^{-1} region as a shoulder to the main bR_{570} peak. This intensity is most probably due to L_{550} or bR_{550}^{DA} (see later in discussion) which will also be resonance enhanced by 514.5 nm excitation. There is very little M_{412} evident despite the fact that excitation with 514.5 nm radiation will cause more efficient pumping of the photocycle than excitation of 488.0 nm and 476.5 nm. This suggests that resonance enhancement of bR_{570}, and possible depopulation of L_{550} and K_{610}, that occurs due to light-induced cycling which bypasses M_{412} (Figure 1), renders negligible the scattering from M_{412}. The strong peak at 1644 cm^{-1} in the Schiff base region also suggests that deprotonation of the Schiff base has not occurred for the majority of scattering bR molecules.

For NBS:bR 10:1 there is a significant increase in the asymmetry of the ethylenic band to the high energy side. This suggests that there is an increase in the population of L_{550}, or of intermediate resembling L_{550} in structure and appearing prior to M_{412}. There are also small changes in the fingerprint region and loss of definition of the protonated Schiff base vibrational band. These trends show up more clearly with blue excitation at 488.0 nm and 476.5 nm. Resonance enhancement of M_{412} with these wave-lengths enables M_{412} to become an important contributor to total scattering intensity, resulting in strong peaks at 1567 cm^{-1} and 1622 cm^{-1}. Again for NBS:bR 8:1, there is considerable increase in scattering for band positions that can normally be attributed to L_{550} or photolysis and relaxation products in the extended reaction scheme (Figure 1). Marcus and Lewis[35] have also identified a precursor to M labelled X which has an ethylenic stretch frequency at 1552 cm^{-1} and so may also be contributing to new scattering intensity in this region. Broadening in the Schiff base vibrational region is also evident.

These results show that the effect of modification of two tryptophans in the bR sequence is to increase the number of conformational variants of retinal contributing to total scattered intensity. The broadening and merging of the Schiff base vibrations at 1644 cm^{-1} and 1622 cm^{-1} indicates that the protonation of the Schiff base is also variable and significantly perturbed from the normal steady state situation.

The changes observed in the CW spectra of NBS:bR 10:1 are consistent with retinal assuming a '13-cis type' configuration rather than the 'all-trans' type which is known to be the isomer present in native bR_{570}.[28] Raman spectra of Schiff bases of retinal in solution show that 13-cis retinal has an ethylenic stretch frequency which is up to 10 cm^{-1} higher in energy than the equivalent all-trans vibration.[25,28,29,30] In the fingerprint region, the 1200 cm^{-1} band of all-trans retinal shifts by about 5 cm^{-1} to higher energy in Schiff bases of 13-cis retinal. The Schiff base vibration for protonated 13-cis retinal is about 3 cm^{-1} to lower energy than in all-trans retinal. It will be recalled that changes in NBS-bR Raman spectra consisted of additional intensity to the high energy side of the 1200 cm^{-1} and ethylenic stretch bands, and to the low energy side of the protonated Schiff base vibration (see Figures 6, 7 and 8).

1.1.2. Pulsed excitation. The steady state flow Raman spectra gives no indication of the temporal characteristics of the species responsible for the band at 1540 cm^{-1}. Time resolution was therefore introduced by using pulsed excitation on a nanosecond timescale. Single and double pulsed techniques were used, the former enabling the first few nanoseconds of the cycle to be investigated, while double pulsed techniques allowed timescales of the order of μs to be probed.

1.2. Experimental Details

Single pulsed Raman spectra were generated using a Molectron Nd:YAG pumped tunable dye laser (5 ns pulse duration, 10 Hz repetition rate) emitting at 514.5 nm (C500 dye, 2.0 mJ/pulse) for comparison with CW spectra obtained with the same excitation wavelength. Melles Griot cut off filters removed Rayleigh scattering. A 50 nm focal length cylindrical lens focussed the excitation beam through the sample (25 μM, aqueous suspension, pH 6) which was flowed continuously from a 7 ml reservoir using a peristaltic pump. A 2.5 m Czerny-Terner scanning spectrometer dispersed the light scattered by the sample at 90° to the laser radiation. Signal from the EMI 9816 QB photomultiplier was averaged with a PAR boxcar signal averager (162/165). Spectra so obtained were calibrated using the Raman band positions of toluene.

Two Molectron MY 34 Nd:YAG lasers were used for the dual pulsed experiments (see Figures 9 and 10). One laser provided a high intensity (7 to 10 mJ/pulse) unfocussed pump pulse while the probe pulse (1 to 15 mJ/pulse) was generated by the second laser at different delay times after the pump and was focussed in the sample. The 532 nm output of one laser was used as the pump. The probe wavelength at 436 nm was generated by Raman shifting the second laser in high pressure H_2 (200 psi) and selecting the first antistokes output at 436 nm. The pulse width of both pump and probe pulses was 15 ns. Raman spectra were collected with 1 to 1.5 mJ/pulse at 436 nm excitation focussed weakly in the sample with a 200 nm spherical lens.

Intermediates occurring earlier in the photocycle were investigated using 532 nm as both pump and probe excitation. A 200 nm lens focussed the probe beam weakly (1.5 mJ/pulse) and recollimated the pump beam (7 mJ/pulse). Pump and probe beams were combined using a 25% reflective dichroic filter.

Spectra were recorded using the same detection apparatus described above, the only difference being that spectra were normalized to probe laser intensity using the A/B normalizing option of the PAR boxcar 162 mainframe. Slits were set at 200 μm, giving a spectral resolution of about 2 cm^{-1}. Melles Griot filters (FCG087 and FCG063) were used to select Raman from Rayleigh scattering.

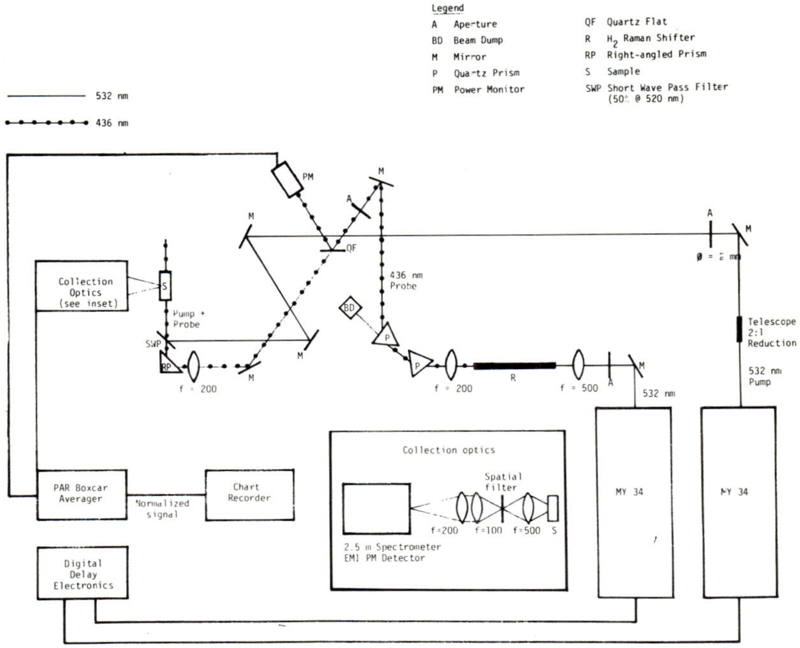

FIGURE 9 Apparatus for pulsed TR^3 spectroscopy of native and modified bR. The pump excitation is provided by the 532 nm output of one Molectron Nd:YAG laser (MY 32) while the probe excitation is produced by Raman shifting the 532 nm output of a second Molectron Nd:YAG laser (MY 34) and selecting the first anti-Stokes emission at 436 nm. Pump and probe beams are combined using a short wave pass dichroic filter (50% transmittance at 520 nm). For sampling and timing details see text.

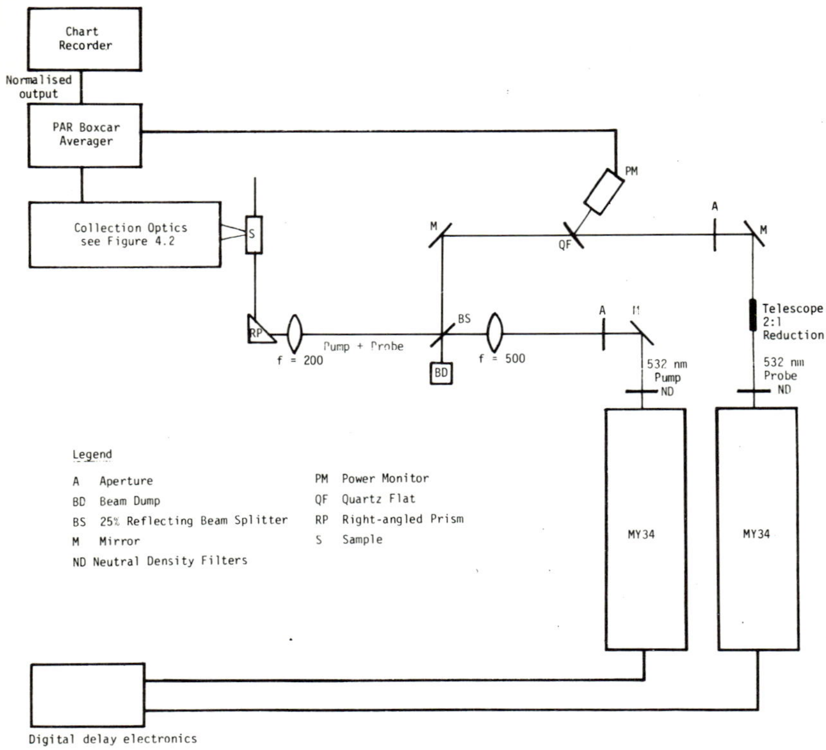

FIGURE 10 Apparatus for pulsed TR^3 with 532 nm pump and probe excitation. The 532 nm output of both Nd:YAG lasers is combined using a 25% reflecting beam splitter. For sampling and timing details see text.

Delays between pump and probe pulses were generated by a home built digital unit (10 ns time resolution) which triggered the laser flashlamps with the appropriate time separation. The Q switch synchronous voltage output from the pump laser was used to initiate an internal clock in a Tektronix 7D11 digital delay plug-in unit mounted in a Tektronix 7834 400 MHz storage oscilloscope.

The output trigger from the delay unit was delayed for the required time span and then used to trigger the Q switch of the probe laser.

Samples (15-20 μM, in 50 mM phosphate buffer, pH 7) were continuously flowed through the 5 ml capacity quartz cuvette from a 30 ml reservoir using a variable flow pump (Micropump P25336). A flow rate of 0.5 ml/sec was used for experiments where delays of milliseconds were required. For shorter delays a flow rate of 1 ml/sec was used. The probe beam (diameter 0.2 mm at focus estimated from the intensity profile of the image on the spectrometer slits) was aligned in the centre of the 2 mm diameter pump beam. With a bulk flow rate of 1 ml/sec, the estimated time for non-photolysed membrane fragments outside the pump volume to diffuse into the probe volume is at minimum, about 12 ms. The sample flow ensured that fragments exposed to laser light in a pump/probe sequence were replaced by unphotolysed fragments for the next sequence since the time between successive pump/probe sequences was 100 ms (10 Hz repetition rate). This method allowed the recording of spectra averaged over approximately 3 h without observable loss of M_{412} yield due to photodegradation. However, NBS-bR samples showed greater susceptibility to photodegradation and were discarded once changes observed in the time-resolved spectra were no longer reversible.

1.2.1. Results and observations. The similarity in the ethylenic region of the Raman spectra of NBS:bR 4:1 and native bR as opposed to NBS:bR 10:1 is shown in Figures 11 and 12 which were generated with pulsed excitation. The temporal pulse width of the laser used to obtain Figure 11 is about 5 ns. Excitation was chosen at 514.5 nm

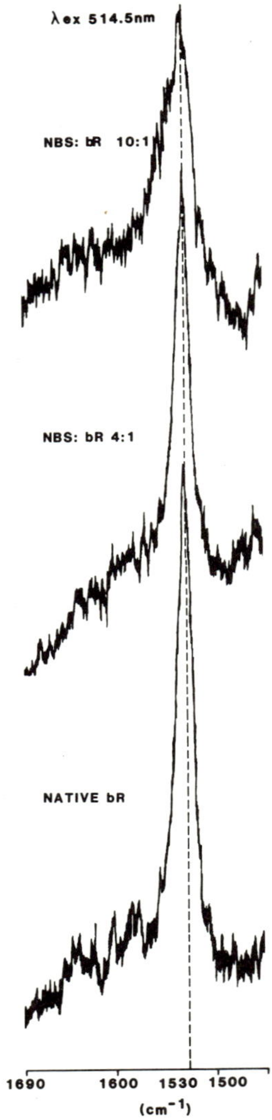

FIGURE 11 Pulsed Raman spectra of native bR, NBS:bR 4:1 and NBS:bR 10:1 using 514.5 nm excitation (2.0 mJ/pulse, 5 ns duration). Spectra were recorded with the same detection sensitivity. Protein concentration, 25 μM; spectral resolution, 2 cm^{-1}.

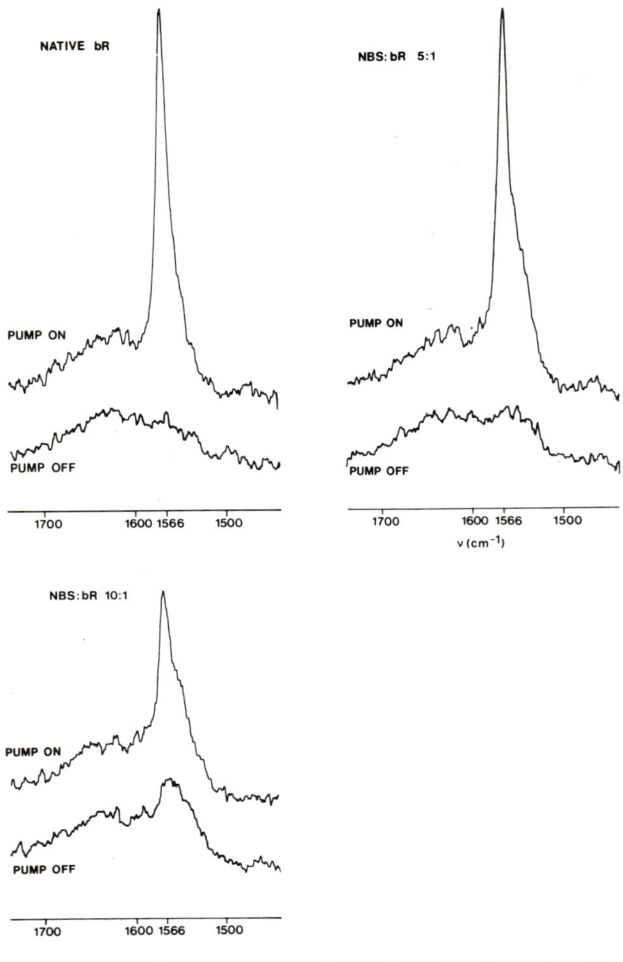

FIGURE 12 TR^3 spectra of native bR, NBS:bR 5:1, and NSB:bR 10:1 using 532 nm pump (10 mJ) and 436 nm (1.5 mJ) probe excitations. Time delay, 200 μs; protein concentration 20 μM in 50 mM sodium phosphate buffer, pH 7; spectral resolution, 2 cm^{-1}. Spectra are reproduced with equivalent vertical enlargement to allow direct comparison of peak intensities.

to allow comparison with the CW Raman spectra obtained with the same excitation wavelength. Additional intensity appears on the high energy side of the 1530 cm^{-1} bR_{570} peak for NBS:bR 10:1, as observed in the CW spectrum. The height of this peak is also considerably diminished with respect to native bR and NBS:bR 4:1.

The spectra of Figure 12 were generated using the dual pulse technique outlined in the experimental section in which a pump pulse at 532 nm was followed after a 200 μs delay by a low intensity probe pulse at 436 nm (15 ns duration). Figure 13 shows that this delay maximizes production of M_{412} for both native bR and NBS:bR 10:1 and that the rise and decay times for M_{412} production are only slightly changed by NBS modification. Again, native bR and NBS:bR 5:1 display very similar spectra in the ethylenic region with a strong peak appearing at 1566 cm^{-1} as expected for M_{412}, whereas the same region for NBS:bR 10:1 is diminished in intensity and considerably broadened towards the low energy side of the 1566 cm^{-1} band. Figure 14 shows comparison of native bR and NBS:bR 10:1 Raman spectra down to 1100 cm^{-1} for a delay of 200 μs between the pump and probe pulses. The broadening in the ethylenic region observed for NBS-bR is associated with considerable loss of definition for the band structure in other regions of the spectrum, as also observed for CW excitation.

Intermediates occurring earlier in the photocycle were investigated using 532 nm excitation for both pump and probe beams. Time-resolved spectra in Figure 15 were recorded with a pump-probe delay of 10 μs at which time the population of L_{550} is calculated to be at a maximum. L_{550} is reported to have an ethylenic stretch frequency near 1540 cm^{-1} (Table 1). Changes described here were

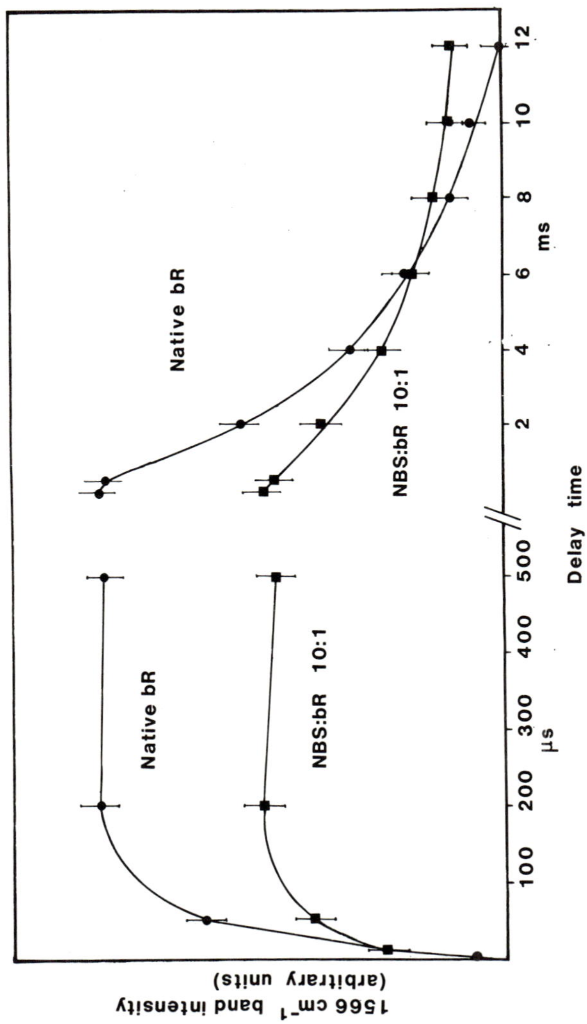

FIGURE 13 The formation and decay of the ethylenic stretch band at 1566 cm^{-1} for M_{412} as measured by peak height for various delays between the 532 nm pump pulse and the 436 nm probe pulse. Protein concentration 15 µM in 50 mM sodium phosphate buffer, pH 7.

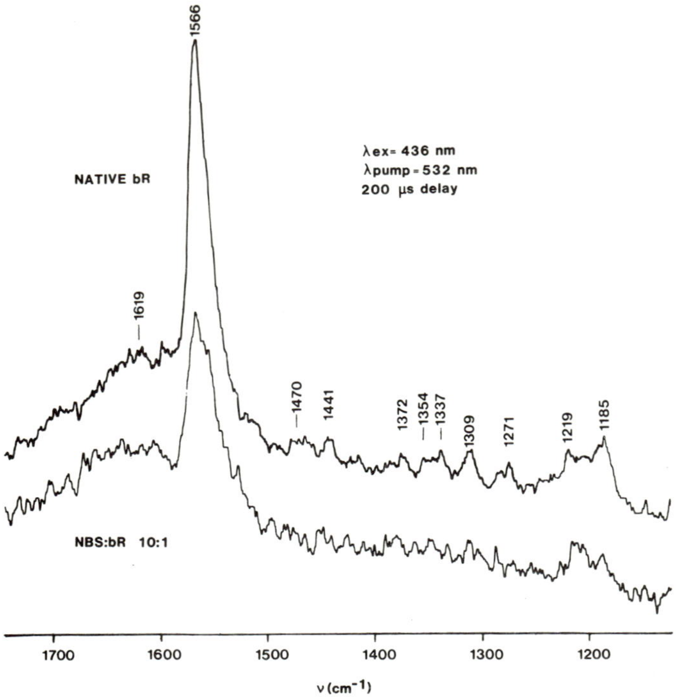

FIGURE 14 Pulsed TR³ spectra of native bR and NBS: bR 10:1 using 436 nm probe excitation (1.5 mJ) and 532 nm pump excitation (10 mJ). Delay, 200 μs; protein concentration, 15 μM in 50 mM phosphate buffer, pH 7: spectral resolution, 2 cm^{-1}.

most apparent using a 10 μs delay. Native bR displays a shoulder appearing at 1540 cm^{-1} for a delay of 10 μs. As expected, most intensity appears at 1530 cm^{-1} (bR_{570}) and 1522 cm^{-1} (K_{610}), consistent with expectations based on the excitation duration of 15 ns. (The effects of photolysis induced intermediates will be discussed in the next section). With a delay of 10 μs the 1522 cm^{-1} band increases in intensity, the 1530 cm^{-1} band is slightly decreased and the shoulder appears at 1540 cm^{-1}.

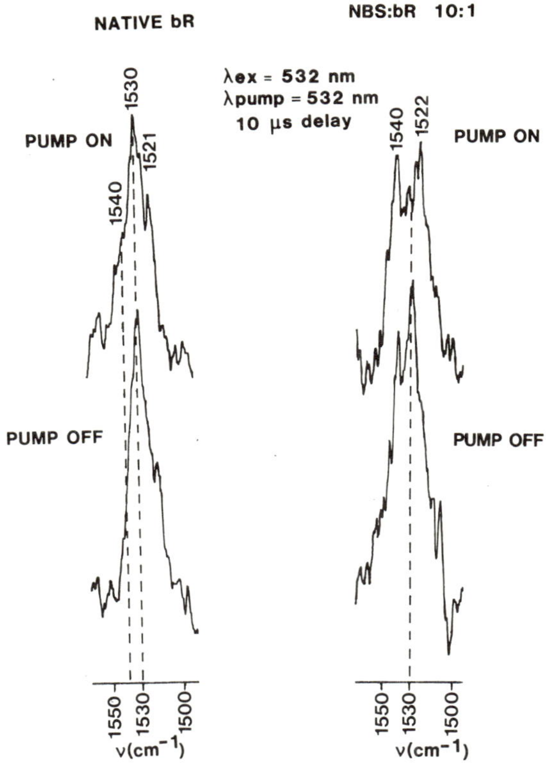

FIGURE 15 Changes induced in the ethylenic vibrational band structure for native bR and NBS:bR 10:1 when a 10 μs delay is set up between the pump (532 nm, 7 mJ) and probe (532 nm, 1.5 mJ) pulses. Protein concentration, 20 μM in 50 mM sodium phosphate buffer, pH 7; spectral resolution, 2 cm^{-1}.

As foreshadowed by CW and pulsed Raman spectra described above, NBS:bR 10:1 displays different cycling characteristics on the microsecond time scale. Without the pump radiation, there is already an intense band at

1540 cm^{-1} present in addition to the main band at 1530 cm^{-1}. There is also some structure on the low energy side of the 1530 cm^{-1} band which is possibly attributable to K_{610}. With the pump on and a 10 μs delay the most intense scattering appears in the 1540 cm^{-1} and 1522 cm^{-1} bands, with the 1530 cm^{-1} band reduced in intensity. The additional band structure in the spectrum of NBS:bR 10:1 results in a band width (fwhm) of more than 30 cm^{-1} in the ethylenic region as compared with 23 cm^{-1} in native bR obtained with the same pump and probe intensities and delay. With the pump off, the band width for the 1530 cm^{-1} peak is 18 cm^{-1} for native bR and 23 cm^{-1} for NBS:bR 10:1.

2. DISCUSSION AND CONCLUSIONS

NBS:bR 5:1 and NBS:bR 10:1 samples may be distinguished by several characteristics. The former, for which only one tryptophan (TRP-10 or TRP-12) is oxidized displays cycling and spectroscopic characteristics which are very similar to those of native bR. It is only when two tryptophans on average per bR molecule are oxidized by increasing the molar ratio to 10:1 that cycling efficiency is affected, and the apparent torsional freedom of retinal about single bonds (ring-chain planarization) is increased.[19]

This trend in distinguishing characteristics between native bR and NBS:bR 5:1 on the one hand, and NBS:bR 10:1 on the other, is also observed in the Raman spectra of these samples presented here. Observable changes to Raman spectra, most noticeable in the ethylenic region occurred only when more than one tryptophan was oxidized in the bR sequence (i.e. for NBS:bR $\geq$ 8:1).

If the spectra of Figure 12 are compared with Figures

7 and 8 which show the steady state M_{412} contribution to Raman scattering for the CW excitation experiments, some interesting observations can be made. Firstly, both CW and pulsed excitation spectra show diminished M_{412} yield for NBS:bR 10:1 (or 8:1) as expected from flash photolysis results[19] and Figure 13. Secondly, the new intensity appearing between 1530 cm^{-1} and 1550 cm^{-1} for NBS:bR in the CW excitation spectra is not apparent in the pulsed excitation spectra. CW excitation at 476.5 nm produces a resolvable band at 1540 cm^{-1} (Figure 8); no such band is observed in Figure 12 obtained with pulsed excitation at 436 nm. The time-dependence of the 1540 cm^{-1} band therefore requires further characterization before an assignment for the origin of this band can be made. While the steady state Raman spectra using 476.5 nm excitation might suggest that the species responsible for the 1540 cm^{-1} band is resonance enhanced by the excitation and appears on a time scale comparable with M_{412} ($\sim$50 μs), the pulsed Raman spectra (Figures 12 and 14) discount the possibility that the new band is due to a new species which is coupled to M_{412} production. Figures 12, 13 and 14 show that while yield of M_{412} is reduced for NBS:bR 10:1, the species of M_{412} produced by photocycling in NBS:bR appears to be similar to that produced by normal photocycling in structure and rise and decay times.

In the description of CW Raman spectra obtained for NBS-bR in the previous section, it was pointed out that the main effect of modification of 2 tryptophans was an increase in the asymmetry of the ethylenic band to the high energy side. Small changes in the fingerprint region and loss of definition of the protonated Schiff base vibrational band were also associated with NBS:bR 10:1 spectra. These changes were consistent with retinal forming a '13-cis type' configuration. The broadening of bands was an indication of more conformational variants of retinal being present in NBS:bR 10:1 samples than in native bR or NBS:bR 5:1.

The appearance of extra intensity in the 1540 cm^{-1} region with 5 ns pulsed excitation (Figure 11) suggests that the photocycle is perturbed at very early stages. Nevertheless, some form of cycling still occurs: a strong Raman peak at 1567 cm^{-1} due to M_{412} appears with CW excitation, and Figure 13 shows that cycle kinetics are not very different from those of native bR. Evidently, the perturbations to the active site do not completely halt the cycle but merely reduce the efficiency of M_{412} production. The appearance of a band in CW spectra at 1540 cm^{-1} that is not clearly resolved in the nanosecond pulsed spectra suggests that the particular protein retinal configuration responsible for the species is produced by pumping of the photocycle, albeit a perturbed cycle. The 1540 cm^{-1} band, therefore, is not representative of a new species of bR relaxed into a state of stasis by blocking of the photocycle.

The probe intensities used in the CW and single or double pulse TR^3 spectra reported here were such that the probability of a particular bR molecule interacting with more than photon was high. Under such conditions, a photostationary equilibrium may exist between bR_{570} and its photoproducts because the quantum yield of formation of bR_{570} from K_{610} is about 0.7.[7,18,31] Successive absorption of photons may also cause the light-induced production of bR_{550}^{DA} (reference 16, Figure 1). Coupling to the 'cis cycle', the cycle initiated by excitation of bR^{DA}_{550} (the trans cycle refers to the normal photocycle of bR^{LA}_{570} in this terminology) may occur via the excited state of bR_{570}, I, or J_{625} and K_{610}, the relaxation products of I. Consequently, pulsed Raman spectra which are generated using sufficiently high light intensities within the absorption band of bR^{LA}_{570} will create a photostationary equilibrium between several intermediates of the cis and trans cycles.[15]

In double pulse experiments, the intensity of the pump beam was chosen such that the yield of M_{412} was maximized but 2 photon processes were unlikely. Flash photolysis observation of M_{412} yield showed that saturation of the cycle, that is when photoinduced reconversion of K to bR prevents further production of M_{412}, occurs for pulse energies around 10 mJ (5×10^{17} photons/cm^2) and bR concentration in the range 10 to 20 μM (∿100 photons per bR molecule). The probe pulse was aligned in the centre of the pump pulse where maximum photocycling yield would be achieved. It is estimated that under such conditions for pump and probe alignment and pulse energy, the probe beam sampled bR near the saturation limit where 20 to 30% bR is photocycling, but back conversion to bR is not occurring significantly.

Assuming that photo-induced reactions to intermediates on the 13-cis cycle do not predominate as a result of the pump pulse, a 10 μs delay should result in the accumulation of observable quantities of L_{550}. Maximal L_{550} population is calculated and observed to occur 10 μs after a photolysis flash with a lifetime of the order of nanoseconds.[12] The 10 μs delay spectrum for native shown in

Figure 15 is consistent with a yield of L_{550} of about 20% of total bR molecules present. There is an observable shoulder at 1540 cm^{-1} which can be attributed to L_{550}. The peak at 1521 cm^{-1} is evidence that the probe beam has photolysed some of the bR_{570} molecules that still form the bulk of the sample with only 20 to 30% cycling initiated by the pump pulse. The height of the 1530 cm^{-1} band is reduced by about 20% and broadened from 17.5 cm^{-1} for the pump off spectrum to 22.5 cm^{-1} (fwhm) for the pump on. This is consistent with the population of transient species by the pump and probe beams reducing the number of bR molecules in the bR_{570} state. The Raman scattered photons for a 10 μs delay will therefore represent scattering from bR_{570}, K_{610}, L_{550} and the thermal and photolysis relaxation products of L_{550} and K_{610}.

The photocycle can return L_{550} to bR_{570} without the formation of M_{412} via either the photo-induced product L' or R_2^{550}, a thermal decay product (see Figure 1). R_2^{550} forms in about 40 μs so for a time resolution of 10 μs should not contribute to Raman intensity.[17] L' is a likely contender for Raman intensity but the features of its Raman spectrum are still not defined precisely[32] since it is so closely coupled to L_{550} formation and decay. Stockburger et al.[17] have observed two weak features at 1570 cm^{-1} and 1580 cm^{-1} which are probably due to L' since they are not observed in a low intensity spectrum of nearly pure L. The absorption maximum of L' is probably near 570 nm, slightly redshifted from that of L at 550 nm. Hence, the ethylenic stretch frequency of L' is expected near 1530 cm^{-1} by the linear correlation between ethylenic stretch frequency and λ max observed by Aton et al.[29] If this is the case, then identification of the spectral characteristics of L' in a spectrum containing a large scattering contribution from bR_{570} is not possible.

2.1. The Origin of the 1540 cm^{-1} Band : A Dark-Adapted Species of NBS-bR?

The absence of the 1540 cm^{-1} peak from the spectra of M-M' for NBS-bR (Figure 14) suggests that the intensity of this band follows kinetics similar to L_{550} in native bR in that the intensity of this peak is greatest within a few microseconds of the pump pulse (Figure 15). However, the appearance of the 1540 cm^{-1} band without the pump pulse in the NBS:bR 10:1 spectrum prevents a definite assignment of this peak to L_{550} because the spectrum of native bR under the same conditions does not show this structure. The 532 nm pump and probe spectra provide additional support to the conclusion drawn on the basis of the pulse Raman spectra of modified bR (Figure 11) that the photocycle is significantly perturbed in the early stages subsequent to photon absorption. The presence of significant intensity at 1540 cm^{-1}, even on a nanosecond timescale suggests that a large proportion of retinal bound to the protein containing two or more modified tryptophans assumes a conformation which resembles closely the chromophore structure of L_{550} or bR^{DA}_{550}.

Stockburger et al.[17], in their extensive Raman study of the bR photocycle, identified several intermediates 'R_i' which absorb in the region 550 to 590 nm and have ethylenic stretch frequencies around 1520 to 1550 cm^{-1} (see Table 1). These intermediates form cycles which bypass normal M_{412} formation (see Figure 1). R_1^{570} (or K^C) and $R_1^{'590}$ (L^C_{610}) form a photo-induced conversion to a 13-cis cycle which populates bR^{DA}_{550}. R^{550}_{2} is a relaxation product of L_{550} that competes with production of M_{412} and returns bR to bR_{570}.

The observed growth of intensity in the 1530 to 1550 cm^{-1} region in CW Raman and pulsed Raman spectra is consistent with increased production of these photolysis induced species. The reduction in yield of M_{412} would therefore be explained by depopulation of L_{550} or K_{610} by increased efficiency in these light driven or thermal relaxation pathways which are competitive with the normal photocycle. The intermediate $R_1^{'590}$ (or L^{C}_{610}) in particular has a long lifetime (37 ns, reference 17) and an ethylenic stretch frequency expected to be in the 1520 cm^{-1} region by the λ max, $\nu_{c=c}$ correlation,[29] which might explain the appearance of a strong band in this region in the 10 μs time resolution spectrum of NBS:bR 10:1. If this is the explanation, then the band observed at 1540 cm^{-1} in CW and pulsed spectra is more likely to be due to the photolysis production of bR^{DA} via the cis cycle, which is facilitated by the effects of NBS modification.

The dependency of the scattering at 1540 cm^{-1} upon excitation wavelength is also consistent with the proposal that the dark-adapted species of bR is formed by the interaction of intense light with NBS:bR 10:1. For pulsed Raman spectra with excitation at 514.5 nm (Figure 11), no resolvable band 1540 cm^{-1} was observed, while for single pulse excitation at 532 nm, there exists a clearly identifiable peak at 1540 cm^{-1}. This suggests that excitation closer to the absorption maximum of bR_{570} is more efficient at producing the species responsible for the 1540 cm^{-1} peak and provides evidence for a light driven pathway for the production of this species. Since a light driven pathway for the conversion of intermediates of the light adapted 'trans cycle' to the dark adapted 'cis' cycle has already been shown to exist[16,17] it is not unreasonable to conclude that the increased intensity of the 1540 cm^{-1} peak in CW and pulsed Raman spectra of NBS: bR 10:1 is due to enhanced production of bR_{550}^{DA}, or an equivalent species containing 13-cis retinal.

To conclude, the loss of M_{412} yield from NBS:bR $\geq$ 8:1 samples may be associated with increased propensity in modified samples to thermally- or light-induced branching in the photocycle. There are a number of ways by which the primary photochemical step in particular can be perturbed, the most likely being coupling of the 'trans cycle' to the '13-cis cycle', or alternatively, by reformation of bR_{570}. The appearance of a band at 1540 cm^{-1} in both CW and pulsed TR^3 spectra suggests that some form of perturbed photocycling is taking place instead of, or in addition to, repopulation of bR_{570}. Possible candidates for the origin of the 1540 cm^{-1} band are bR_{550}^{DA} and other intermediates of the cis cycle, of L_{550} itself. However, 'trapping' of L_{550} is not consistent with the observations of a 1540 cm^{-1} peak on a nanosecond timescale as well as in the steady state CW Raman spectrum. Flash photolysis[19] and pulsed TR^3 also indicate the decay of L_{550} (or formation of M_{412}) is not changed enough to account for the large increase in L_{550} population required to account for a strong band at 1540 cm^{-1}. bR_{550}^{DA}, or its cis cycle precursors, are more likely contenders for the new intensity in the ethylenic region. The 1540 cm^{-1} is therefore tentatively assigned to bR_{550}^{DA}, which contains retinal in a 13-cis conformation. Sperling et al.[16]

report that bR^{DA}_{550} can be populated by photolysis of K. It will also be recalled that CW Raman spectra allowed the conclusion that retinal in NBS-bR assumed more 13-cis character than in native bR.

Significant change to photocycle characteristics also requires modification of two tryptophan residues per bR molecule. It is therefore possible that the second tryptophan accessed plays a more important role in the maintenance of optimal active site configuration for activation of the proton pump. However, there is no evidence as yet that NBS reacts with a specific tryptophan after the first residue is modified.[19] A more reasonable explanation for the loss of yield is that the protein structural changes induced by modification of one tryptophan are insufficient to cause measurable changes to the photocycle as monitored by flash photolysis or TR^3. Modification of two or more tryptophans overcomes the threshold in protein stability to environmental change and the photocycle is correspondingly perturbed. Assuming that the effect of oxidation of tryptophan residues in a predominantly α-helix structured backbone is to cause a small amount of unfolding,[19,33,34] it is also possible that the second tryptophan(s) accessed is located such that the small changes in secondary or tertiary structure have a greater effect on the retinal binding site.

One of the primary aims of this work which involves the combination of time resolved spectroscopic studies with protein chemistry was to investigate whether experimental evidence for an interaction between retinal and tryptophan as suggested by the point-charge models[8,13] could be obtained. While such an interaction cannot be totally discounted, the evidence reported here and in a previous paper[19] suggests that if one or more tryptophans are involved in charge stabilization of the retinyl chromophore, then they are not accessible to the NBS reaction at the relatively low concentrations of NBS used in this work. Indeed, oxidation of a residue in the immediate environment of retinal and participating actively in stabilization of electron delocalization and retinal excited state structure would be expected to have a more profound effect on its spectroscopic and photocycle characteristics.

We favour the conclusion that tryptophan residues appear to play a role in maintaining overall protein conformation and rigidity in the protein. The Raman spectra of NBS:bR show the consequences for protein activity when only very small changes in α-helix content and rigidity are induced. The quantum yields of competitive reactions increase to the extent that the yield of the productive proton pumping cycle can be decreased by anything up to 70% for the same intensity conditions. This provides evidence for the maintenance of protein rigidity in the membrane environment as a necessary prerequisite to mediation of binding site conformation and protection of the binding site from hydrolysis reactions. Most importantly, the efficiency of the protein in specifically channeling the reaction pathway of excited state retinal to conformational changes that induce the correct protein conformational changes for proton pumping, has evidently been decreased by NBS modification. In addition, the protein structured changes probably diminish the efficiency with which configurational changes induced in retinal by light excitation can be transmitted by the protein.

REFERENCES

1. W. STOECKENIUM, in Photosynthetic Bacteria, ed. R. K. Clayton (Plenum Press), 571-592, (1972).
2. M. OTTOLENGHI, Advances in Photochemistry, 12, 97-200 (1980).
3. R. R. BIRGE, Ann. Rev. Biophys. Bioeng., 10, 315-354 (1981).
4. A. LEWIS, Methods in Enzymology, 88, 561-617 (1982).
5. M. OTTOLENGHI, Methods in Enzymology, 88, 470-491 (1982).
6. W. SPERLING, P. CARL, C. N. RAFFERTY and N. A. DENCHER, Biophys. Struct. Mech., 3, 79-94 (1977).
7. C. R. GOLDSCHMIDT, O. KALISKY, T. ROSENFELD and M. OTTOLENGHI, Biophys. J., 17, 179-183 (1977).
8. K. NAKANISHI, V. BALOGH-NAIR, M. ARNABOLDI, K. TSUJIMOTO and B. HONIG, J. Am. Chem. Soc., 102, 7945-7947 (1980).
9. T. KONISHI and L. PACKER, FEBS Letts., 79, 369-373 (1977).
10. C. N. RAFFERTY, Photochem. Photobiol., 29, 109-120 (1978).
11. B. HESS and D. KUSCHMITZ, FEBS Letts., 100, 334-340 (1979).
12. J. M. FUKUMOTO, W. D. HOPEWELL, B. KARVALY and M. A. EL-SAYED, Proc. Natl. Acad. Sci. U.S.A., 78, 252-255 (1981).
13. B. HONIG, A. D. GREENBERG, U. DINUR and T. G. EBREY, Biochem., 15, 4593-4599 (1976).
14. R. MATHIES and L. STRYER, Proc. Natl. Acad. Sci. U.S.A., 73, 2169-2173 (1976).
15. R. H. LOZIER, W. NIEDERBERGER, M. OTTOLENGHI, G. SIVORINOVSKY and W. STOECKENIUS, in Energetics and Structure of Halophilic Microorganisms, eds. S. R. Caplan and M. Ginzberg (Elsevier/North Holland) 123-139 (1978).
16. W. SPERLING, C. N. RAFFERTY, K.-D. KOHL and N. A. DENCHER, FEBS Letts., 97, 129-132 (1979).
17. M. STOCKBURGER, W. KLUSMANN, H. GATTERMAN, G. MASSIG and R. PETERS, Biochem., 18, 4886-4900 (1979).
18. J. B. HURLEY and T. G. EBREY, Biophys. J., 22, 49-66 (1978).
19. J. P. HARDY, A. E. W. KNIGHT, K. P. GHIGGINO, T. A. SMITH and P. J. ROGERS, Photochem. Photobiol., 39, 81-88 (1984).
20. D. M. ENGELMAN, R. HENDERSON, A. D. McLACHLAN and B. A. WALLACE, Proc. Natl. Acad. Sci. U.S.A., 77,

2023-2027 (1980).
21. D. A. ARGARD and R. M. STROUD, Biophys. J., 37, 589-602 (1982).
22. G. F. KING, P. C. MOWERY, W. STOECKENIUS, H. L. CRESPI and B. P. SCHOENBORN, Proc. Natl. Acad. Sci. U.S.A., 77, 4726-4730 (1980).
23. A. LEWIS, J. SPOONHOWER, R. A. BOGOMOLNI, R. H. LOZIER and W. STOECKENIUS, Proc. Natl. Acad. Sci. U.S.A., 71, 4462-4466 (1974).
24. M. A. MARCUS and A. LEWIS, Science, 195, 1328-1330 (1977).
25. M. BRAIMAN and R. MATHIES, Proc. Natl. Acad. Sci. U.S.A., 79, 403-407 (1982).
26. J. PANDE, R. H. CALLENDER and T. G. EBREY, Proc. Natl. Acad. Sci. U.S.A., 78, 7379-7382 (1981).
27. J. TERNER, C.-L. HSIEH, A. R. BURNS and M. A. EL-SAYED, Proc. Natl. Acad. Sci. U.S.A., 76, 3046-3050 (1979).
28. M. J. PETTEI, A. P. YUDD, K. NAKANISHI, R. HENSELMANN and W. STOECKENIUS, Biochem., 16, 1955-1959 (1977).
29. B. ATON, A. G. DOUKAS, R. H. CALLENDER, B. BECHER and T. G. EBREY, Biochem., 16, 2995-2999 (1977).
30. R. E. COOKINGHAM, A. LEWIS and A. T. LEMLEY, Biochem., 17, 4699-4711 (1978).
31. C. R. GOLDSCHMIDT, M. OTTOLENGHI and R. KORENSTEIN, Biophys. J., 16, 839-843 (1976).
32. J. B. HURLEY, B. BECHER and T. G. EBREY, Nature, 272, 87-88 (1978).
33. M. Z. ATASSI, M. T. PERLSTEIN and A. F. S. A. HABEEB, J. Biol. Chem., 246, 3291-3296 (1971).
34. V. I. TEICHBERG, C. M. KAY and N. SHARON, Eur. J. Biochem., 16, 55-59 (1970).
35. M. A. MARCUS and A. LEWIS, Biochem., 17, 4722-4735 (1978).
36. J. TERNER, C.-L. HSIEH and M. A. EL-SAYED, Biophys. J., 26, 527-541 (1979).

CHAPTER 7

TIME-RESOLVED RESONANCE RAMAN SPECTROSCOPY OF INTERMEDIATES IN THE BACTERIORHODOPSIN PHOTOCYCLE: DIRECT PHOTOCONVERSION OF BR_{570} TO M AND M'

I. GRIEGER and G.H. ATKINSON,
Departments of Chemistry and Optical Sciences,
University of Arizona,
Tucson,
Arizona 85721.

1. INTRODUCTION

The photocycle of bacteriorhodopsin (BR) has been recognized for more than 13 years to control the translocation of protons across the purple membrane of *Halobacterium halobium*.[1] More recently, the BR photocycle has been used extensively as a biophysical model for visual processes involving the chromophore retinal.[2] It is both contexts that resonance Raman spectroscopy has been utilized to study the molecular structures and the dynamics associated with these molecular transformations as they are reflected in the vibrational degrees of freedom of the reaction intermediates. The BR photocycle is initiated photolytically by excitation into the absorption band of the retinal chromophore which has a maximum near 570 nm.[3,4] Optical spectroscopies have been used to identify several molecular intermediates formed by thermal reactions subsequent to BR_{570} excitation. In order to study these thermal intermediates, the radiation near 570 nm used to initiate the photocycle ideally should

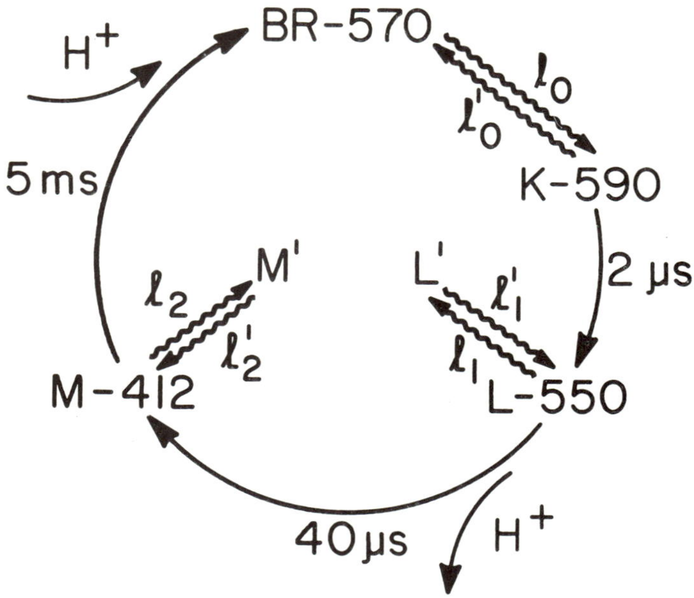

FIGURE 1 Photocycle of bacteriorhodopsin (BR). The ℓ_o and ℓ_o' terms denote photoinduced rate coefficients for photoconversion. Reaction rates refer to PM suspensions at neutral pH and room temperature. The intermediates are given with a subscript to indicate the wavelength of their absorption maximum. The sequence BR-570 $\rightleftarrows$ K-590 → L-550 → M-412 → BR-570 is defined as thermal photocycle (see text).

be of a wavelength and duration so as to insure that only BR_{570} is excited. Under these circumstances, the intermediates in the BR cycle would be formed only by thermal reactions.

This ideal excitation condition is exceedingly difficult to achieve in practice, however, as illustrated in the schematic drawings in Figure 2. The initial

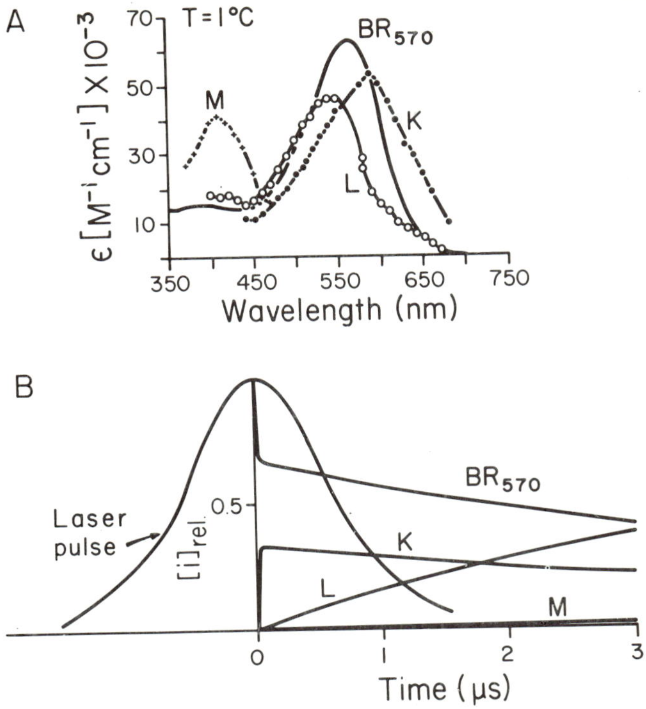

FIGURE 2 (a) Absorption spectra of the intermediates of the BR photocycle at 1^oC.[13] (B) Time evolution of the intermediates in the microsecond time regime compared with a 1 μs laser pulse. The relative concentrations of BR_{570}, K, L, and M ($[i]_{rel}$) are shown schematically.

intermediates, K and L, have absorption spectra which strongly overlap that of BR_{570} itself. Thus, the wavelength of the excitation cannot be chosen to selectively pump only BR_{570}. In addition, both K and L are formed rapidly (e.g., K in < 10 ps) after BR_{570} excitation (Figure 2b). The duration of the radiation, therefore, must be of a few picoseconds in order to time resolve only

BR_{570} excitation. In previous work, the radiation used to initiate the photocycle often had durations of 1 μs (FWHM) (as in Figure 2b) or longer and therefore, BR_{570}, K, and L were all pumped into excited electronic states. The photochemistry of K and L were thereby superimposed on that of BR_{570} itself. Excitation conditions such as these photolytically interrupt the BR photocycle by introducing new reactions involving the excited electronic states of the intermediates themselves (e.g., K* and L*, etc.). These photochemical reactions can obscure the basic thermal cycle arising from BR_{570}, the chemistry which is often of prime interest.

In the work presented here, the excitation conditions used to initiate the BR photocycle from BR_{570} are examined in order to (1) establish when photostationary mixtures of BR_{570} and K are present and (2) search for photolytic interruptions of the BR photocycle involving M and M' (Figure 1). Both optical absorption spectroscopy and time-resolved resonance Raman (TR^3) spectroscopy are used to monitor the effects of BR_{570} excitation.

2. EXPERIMENTAL

(a) Transient absorption experiment. Transient absorption experiments were carried out with a N_2 pumped dye laser (Lambda Physics, Model M2000 and FL 2000T) having a pulse width of 5 ns (GWHM) at 514.5 nm. The 5.4.5 nm laser pulse was used to initiate the photocycle and to monitor the transient absorption of the BR sample. The optical density of the sample was measured directly from the transmitted laser intensity by using a laser photometer (Lambda Physics, Model LF 300).

(b) TR^3 experiment. The instrumentation used to obtain TR^3 spectra is shown in Figure 3. The experiments were performed in a pump-probe configuration utilizing two separate pulsed lasers (a Nd:YAG laser and a Nd:YAG pumped dye laser). Both laser systems were electronically synchronized by a variable time delay (Δt). The second harmonic of the YAG laser at 532.0 nm was used to initiate the BR photocycle (Figure 1). A second laser operating at 420 nm with a Δt = 200 μs generated resonance Raman scattering from the M (and M') intermediate*. The pulse width for both lasers were 10 ns (FWHM) and both were operated at 15 Hz. The two laser beams were focussed colinearly into the sample (Figure 3) and in order to obtain the fresh sample condition, the spinning cell technique was used.[5] The Raman scattering was collected at 90° to the incident radiation and focussed onto a 300 μm slit of a triple monochromator (Spex Triplemate) equipped with a 2400 groove/mm holographic grating. The Raman scattering was detected by a double intensified vidicon camera (EG & G, PAR optical multichannel analyzer, Model 12051 ISIT).

The detection system was interfaced to an on-site computer for data acquisition and analysis. TR^3 spectra of the C = C stretching vibration of the M (and M') intermediate were recorded and their relative intensities were used for the kinetic analysis described here.

*Excitation at 420 nm is also known to initiate the photoconversion of M to a new intermediate designated M'. Recording the TR^3 spectra of M with 420 nm radiation, therefore, also establishes a (photostationary) mixture of M and M'.[5]

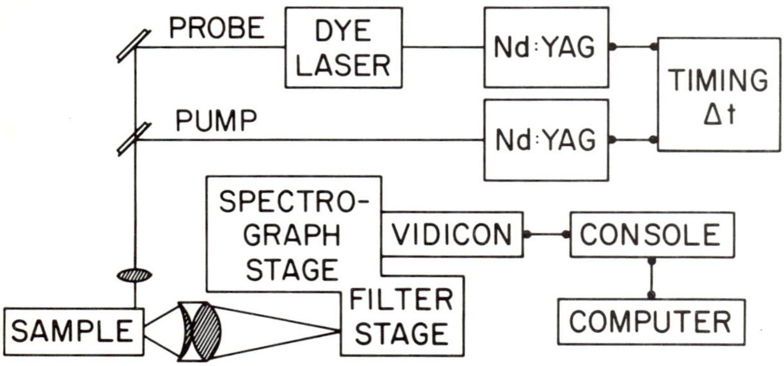

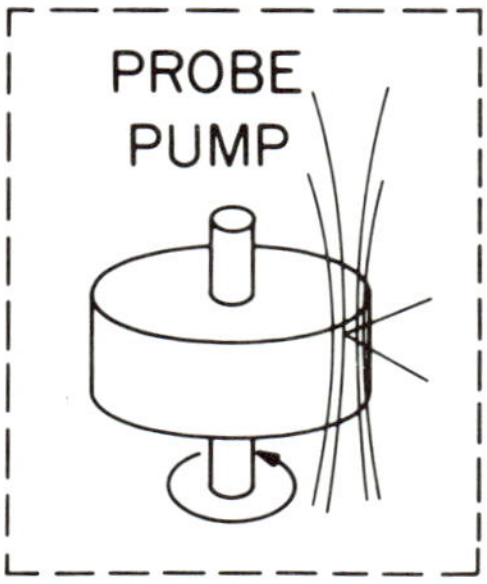

FIGURE 3 Time-resolved resonance Raman (TR^3) instrumentation. The pump laser is a frequency doubled Nd:YAG laser (532 nm) while a Nd:YAG pumped dye laser (420 nm) is used to probe the intermediates via TR^3 spectroscopy. A doubly-intensified vidicon camera is used to detect the resonance Raman scattering. The sample was contained in a spinning cell. Resonance Raman scattering from only the top 0.8 mm of the sample cell was detected. (See text for details).

(c) Material. Purple membrane was prepared from *Halobacterium halobium* S9 strain using essentially the same procedure as Oesterhelt and Stoeckenius.[6] Samples for TR^3 spectra were suspended in double distilled water (pH = 6.6) with an optical density of 4 at λ = 565 nm. The TR^3 spectra were carried out at room temperature.

3. RESULTS

The photostationary interconversion of BR_{570} and K has been shown by a variety of measurements[7-9] to occupy the initial phase of the BR photocycle. This photostationary mixture (PSM) of BR_{570} and K is formed within < 10 ps and thus, is present throughout the 5-10 ns duration of laser pulses used to initiate the BR photocycle in this work. The presence of K has been suggested from the transient absorption spectroscopy.[7,10] The establishment of the BR_{570}/K PSM depends on the excitation energy delivered to the BR_{570} (i.e. excitation power density ≡ EPD) which is correlated with the rate of photoconversion between BR_{570} and K. This is expressed in the term $\overline{n}_o t$ as defined by previous studies[5] where $\overline{n}_o \propto \ell_o$ is the rate of photoexcitation and t the duration of the laser pulse. The value of ℓ_o, therefore, can be controlled readily by the energy per pulse in a laser excitation of BR_{570}. If the same laser wavelength is used to monitor the relative concentrations of BR_{570} and K, then the establishment of the BR_{570}/K PSM can be monitored as a function of the power density in the excitation process. Using pulsed laser excitation (5 ns) at 514.5 nm, these data are shown in Figure 4. Since K absorbs 514.5 nm less strongly than BR_{570} (Figure 2), the shift of the relative concentrations of BR_{570} to K can be seen readily by the decrease in the

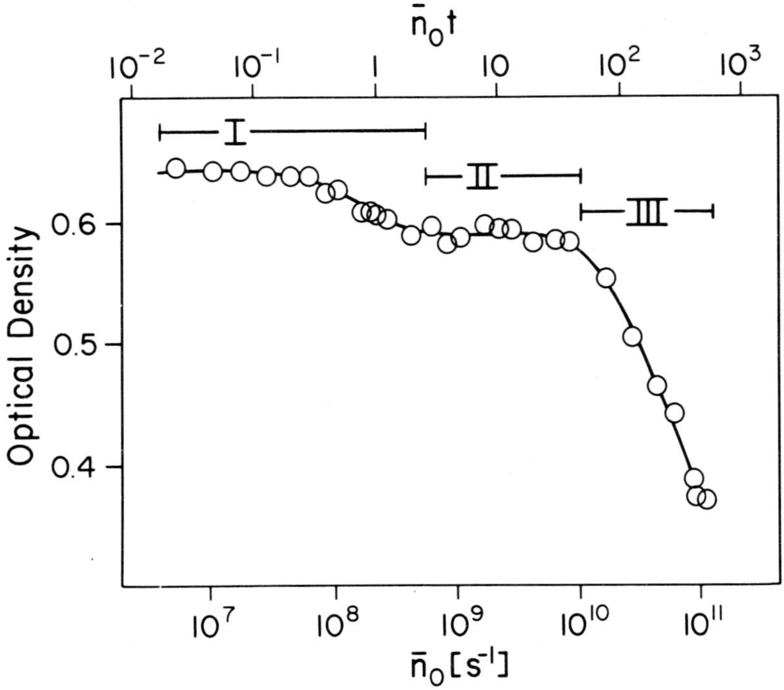

FIGURE 4 Transient absorption at 514.5 nm vs. the amount of photoexcitation, (i.e. $\bar{n}_0 t$). See the text for descriptions of regions I, II, and III.

absorbance of the sample at 514.5 nm as EPD increases (Figure 4). In the higher $\bar{n}_0 t$ values of region I where the OD decreased with EPD, the PSM has yet to reach a stable value. In the plateau region (II), the EPD is sufficiently high to insure the existence of a PSM which has stabilized the relative concentrations of BR_{570} and K.

A second phenomenon occurs when the EPD is increased further. Above $\bar{n}_0 t$ values of 50, the OD of the sample at 514.5 nm starts to decrease again and continues to decrease for $\bar{n}_0 t$ values as high as 10^3 (Figure 4). Since the BR_{570}, K, and L all absorb at 514.5 nm (Figure 2), the decreasing

OD in region III suggests the decreasing concentration of all three species in the sample and the formation of an intermediate(s) that does not absorb in this wavelength region. M, absorbing at 412 nm (Figure 2), is a prime candidate for such a product. This conclusion remained circumstantial, however, since in these experiments no product was directly observed.

From a general viewpoint, the high EPD in region III photolytically interrupts the BR photocycle by introducing a new photochemical channel. The 5 ns duration of the excitation laser pulse leads one to conclude that this new channel originates from excited states of either BR_{570} or I or both, but not L since very little of L has been formed within 5 ns (Figure 2).

The photochemistry of BR_{570} for EPD values in region III (Figure 4) also can be observed clearly from TR^3 spectroscopy. TR^3 spectroscopy was used to reveal the resonance Raman (RR) bands for M[5,6] and recently M'.[11] These spectra provide the opportunity to observe the formation of M and M' directly from TR^3 spectroscopy. The C=C stretching frequencies of M and M' near 1567 cm^{-1} and 1561 cm^{-1}, respectively, were used to examine the EPD relationship with BR_{570}. The relative intensity of the 1567 cm^{-1} and 1561 cm^{-1} pair of RR bands was used to measure the relative concentrations of both M and M'. The M and M' concentration 200 μs after 532 nm excitation of BR_{570} is shown in Figure 5 as a function of the 532 nm EPD. The M and M' intermediates observed in these experiments were formed from BR_{570} via the thermal photocycle route. The amount of M and M' was found to increase as the PSM established stable concentrations of BR_{570} and

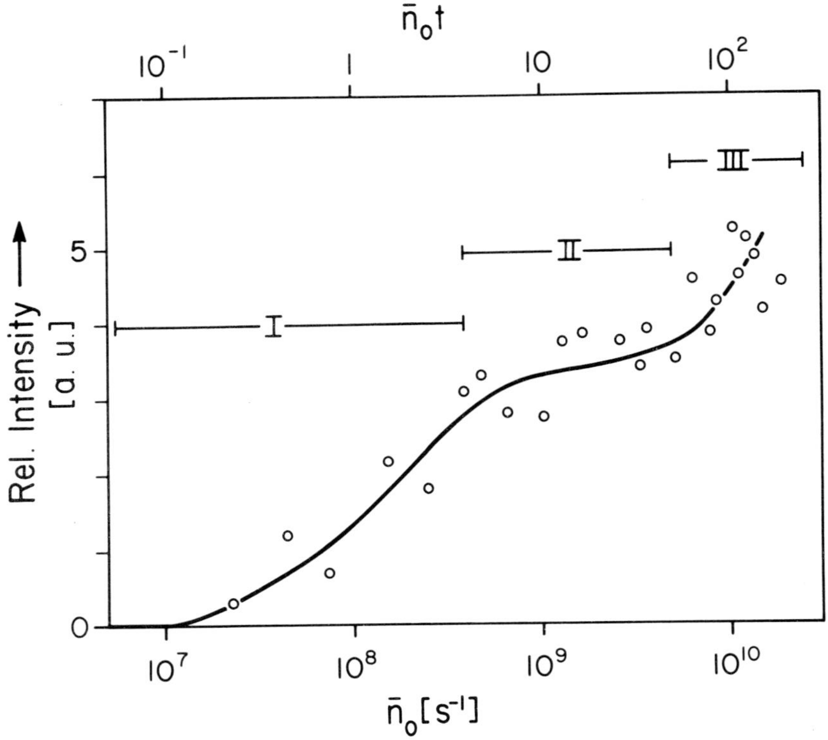

FIGURE 5 Relative intensities of the TR^3 bands of the C=C stretching vibration for M and M' vs. the amount of photoexcitation ($\bar{n}_0 t$) of BR_{570} at 532 nm. The time delay between the pump and probe laser pulses is 200 μs. The laser used to generate resonance Raman scattering from M and M' was operated at 420 nm. The regions I, II and III correspond to the same regions shown in Figure 4.

K. Regions I and II in Figure 5 correspond to the same EPD values observed in regions I and II in Figure 4. Thus, M and M' are formed via the thermal route passing from BR_{570} and K through L. At ℓ_o values approaching region III, however, the relative concentrations of M and M' again increase. Thus, region III in Figure 5 corresponds closely with region III in Figure 4 which monitored the destruction of the BR_{570}/K PSM. These strong correlations suggest that M and M' are produced directly from BR_{570} without proceeding via the thermal route (i.e. without involving L).

4. DISCUSSION

The photolytic conditions under which the PSM of BR_{570} and K is established and the relative importance of the L and M intermediates can be derived from the data presented in Figures 4 and 5. These data need to be viewed in the context of the absorption spectra and kinetic properties of BR_{570}, K, L, and M (Figure 2).

The change in the OD of the BR sample at 514.5 nm in regions I and II of Figure 4 reflects the establishment of the BR_{570}/K PSM. The M (and M') intermediates cannot contribute to the optical absorption at 514.5 nm. Since the L intermediate absorbs 514.5 nm radiation to approximately the same degree as BR_{570} (Figure 2), changes in the relative concentration of L cannot significantly affect the 514.5 nm absorbance of the sample. Thus, neither M (and M') nor L can contribute to the decrease in the OD of the sample observed in regions I and II (Figure 4). The K intermediate, however, absorbs at 514.5 nm less strongly than BR_{570} (and L and as the relative concentrations of the BR_{570}/K PSM shift toward K, one anticipates a decrease in the OD at 514.5 nm. It is the establishment of the PSM between BR_{570} and K that is observed for larger values of $\bar{n}_0 t$ in region I and throughout region II of Figure 4. The plateau in region II reflects the fact that the relative concentrations of BR_{570} and K are stabilized through their photolytically-driven interconversion. Increases in the EPD in region II, therefore, simply maintain the PSM. This interpretation, derived from a consideration of absorption wavelengths, is completely consistent with the time dependent behavior of BR_{570}, K, L, and M. Since the OD of the sample is measured only over the 5 ns pulsewidth of the 514.5 nm laser, only BR_{570} and K could contribute significantly to the data presented in Figure 4.

The decrease in the OD of the sample at 514.5 nm observed in region III ov Figure 4 signals the onset of a major perturbation in the BR_{570}/K PSM, the cause of which can be found from a consideration of the other intermediates in the BR photocycle. The increased formation of L can be eliminated from consideration since it absorbs 514.5 nm more strongly than K. The formation of M, however, could account for this decrease in OD since M (and M') do not absorb at 514.5 nm (Figure 2). It is again important, however, to recall the 5 ns time interval over which the OD measurement is made. In the thermal BR photocycle (Figure 1), M is formed from L over periods of 40 μs or longer and therefore, iz could not contribute to the OD change observed in 5 ns if this thermal route was the only one leading to M (and M'). These OD measurements suggest that at high EPD values such as those in region III (Figure 4), a new photoconversion channel opens which forms M directly from the BR_{570}/K PSM without proceeding through L. The time scale of this new channel would be in the 5 ns regime.

Another view of the photoconversion of BR_{570} is contained in the TR^3 spectroscopy of M (and M') obtained under the same excitation conditions. The resonance Raman (RR) spectrum of M, previously reported from time-resolved measurements,[5] contains a prominent feature (C=C stretching frequency) at 1567 cm^{-1}. The RR spectrum of M', recently recorded via TR^3 spectroscopy,[11] contains a strong C=C stretching band at 1561 cm^{-1}. The relative intensities of these two RR bands are photolytically coupled because of the photoconversion between M and M'.[12] The combined intensities of the 1561 cm^{-1} and 1567 cm^{-1} bands can be used to monitor the relative concentrations of M and M'. If the energy of the probing radiation used to generate TR^3 spectra (i.e. 420 nm) is held constant, then the relationship between the excitation conditions of BR_{570} and the formation of M and M' can be accurately monitored. The data in Figure 5 demonstrate that relative concentrations of M and M' depend on the EPD used to initiate the BR photocycle in the same qualitative way observed for the 514.5 nm optical density measurements (Figure 4).

The 200 μs time delay between 532 nm excitation (10 ns FWHM) and 420 nm probing (10 ns FWHM) insures that the M and M' species are derived from the thermal photocycle route (i.e. via L) in regions I and II. The rise in the M and M' concentrations (region I) followed by a plateau (region II) parallels very closely the behavior observed in Figure 4. The decreasing OD at 514.5 nm is matched by increasing M and M' concentrations for the same $\overline{n}_o$ values. The plateau regions also appear for the same values of $\overline{n}_o$.

These results are consistent with the expectation that M and M' will be formed on the 200 μs time scale by proceeding through L.

A new mechanism for the formation of M and M' is signaled by the TR^3 data obtained for $\overline{n}_o t$ values greater than 50 (i.e. region III). The relative concentrations of M and M' increase dramatically as $\overline{n}_o$ reaches region III (Figure 5). This is the same threshold observed in the OD measurements. The 200 μs delay used in these TR^3 measurements demonstrates also that the M and M' intermediates are present 200 μs after 532 nm excitation, but these TR^3 data do not resolve how quickly M and M' are formed from the BR_{570}/K PSM by this new photoconversion channel. Since the decay of M occurs over periods of 5 ms or longer, its rapid formation even during the excitation pulse of 10 ns could still result in presence of M and M' 200 μs later.

The correlation of the OD and TR^3 data, however, does provide a more detailed view. If the similarity in the dependence of OD and TR^3 data on the values of $\overline{n}_o$ (i.e. EPD) are taken to mean that the same molecular process is being monitored, then it can be concluded that M and M' are formed directly from the BR_{570}/K PSM without passing through L. The OD measurements were made over a 5 ns time interval and therefore, in region III, the suggested formation of M and M' must occur without proceeding through L and the thermal cycle (Figure 1). The TR^3 data confirm that M and M' are formed and that their relative concentrations increase for the EPD (i.e. values of $\overline{n}_o$) found in region III.

5. SUMMARY

The excitation conditions (in terms of $\overline{n}_o$) for which a PSM or BR_{570} and K is established and maintained have been defined by both the OD at 514.5 nm and the TR^3 bands of M and M' to be in region I and II of Figures 4 and 5. For these excitation conditions, the thermal photocycle described in Figure 1 is the principle route by which M and M' are formed. When $\overline{n}_o t$ exceeds 50, however, the direct photoconversion of BR_{570} to M and M' becomes efficient and thereby photolytically interrupts the normal BR photocycle. Thus, there are two routes leading to the formation of M and M' from BR_{570}, one over a 40 μs time scale (via L) and one by direct photoconversion in < 10 ns. On the basis of the frequency positions of the C=C bands recorded by TR^3 spectroscopy, the M and M' intermediates formed by these two different channels appear to be the same.

The rapid channel for M and M' formation may be derived from a multiphoton absorption in BR_{570} since the EPD at $\overline{n}_o t$ values above 50 should be sufficient to produce nonlinear absorption. Whatever the mechanism, it is clear that the results of previous studies, as well as the design for future work in BR, should be viewed carefully with respect to the EPD used to initiate the photocycle. The data presented here demonstrate that photolytic interruption of the thermal cycle can readily occur and introduce new photochemical channels originating in the excited electronic states of either BR_{570} or the thermal intermediates of the photocycle.

REFERENCES

1. D. OESTERHELT and W. STOECKENIUS, Nature New Biol., 223, 149 (1971).
2. W. STOECKENIUS, R.H. LOZIER and R.A. BOGOMOLNI, BBA, 505, 215 (1979).
3. D. OESTERHELT, M. MEENTZEN and L. SCHUHMANN, Eur. J. Biochem., 40, 543 (1973).
4. A. LEWIS, J. SPOONHOWER, R.A. BOGOMOLNI, R.H. LOZIER and W. STOECKENIUS, Proc. Natl. Acad. Sci. USA, 71, 4462 (1974).
5. M. STOCKBURGER, W. KLUSMANN, H. GATTERMANN, G. MASSIG and R. PETERS, Biochemistry, 18, 4886 (1979).
6. D. OESTERHELT and W. STOECKENIUS, in Methods in Enzymology, Vol. 31, 667 (1974).
7. M.L. APPLEBURY, K.S. PETERS and P.M. RENTZEPIS, Biophys. J., 23, 375 (1978).
8. T. GILLBRO and V. SUNDSTRÖM, Photochem. and Photobiol., 37, 445 (1983).
9. R.R. BIRGE and T.M. COOPER, Biophys. J., 42, 61 (1983).
10. I. GRIEGER, Dissertation, Georg - August - Universität Göttingen, Göttingen, (1981).
11. I. GRIEGER and G.H. ATKINSON, manuscript in preparation.
12. J.B. HURLEY, B. BECHER and T.G. EBREY, Nature, 272, 87 (1978).
13. R.H. LOZIER, R.A. BOGOMOLNI and W. STOECKENIUS, Biophys. J., 15, 955 (1975).

CHAPTER 8

STRUCTURAL AND KINETIC STUDIES OF BACTERIORHODOPSIN BY RESONANCE RAMAN SPECTROSCOPY

M. STOCKBURGER
Max-Planck-Institut für biophysikalische Chemie,
Göttingen,
(FRG).

Bacteriorhodopsin (BR) is a retinal-binding protein in the cell membrane of Halobacterium halobium. In its chromophoric center retinal is bound to the protein via a Schiff's base (SB) linkage. On illumination the initial chromophoric state, BR-570, runs through a cyclic process which controls BR's function as a light-induced proton pump (Stocckenius et al., 1979).

hν 2 μs

BR_{570} K_{590} L_{550}

H^+ 40 μs H^+

5 ms

M_{412}

(The numbers designate the wavelength of maximum absorption of each chromophore. The thermal decay constants refer to room temperature).

Resonance Raman (RR) spectroscopy is an ideal tool to perform structural and kinetic studies on BR-570 and the intermediates of its photocycle in time-resolved experiments (Stockburger et al., 1979). From the vibrational structure of the RR spectra it thus could be concluded that the SB group is protonated in BR-570 and L-550 but deprotonated in M-412. This implies that at the L-550 stage of the cycle a proton is removed from the chromophore and later at the M-412 stage a proton is added to reconstitute the original state BR-570. From spectroscopic evidence it could be further concluded that water molecules at the chromophoric center stabilize the positively charged SB group and its negative carboxylate counterion (Hildebrandt and Stockburger, 1984).

We have designed kinetic RR experiments to study quantitatively the time-behaviour of BR-570 and its intermediates L-550 and M-412 during the photocycle (Alshuth et al., 1984). The strongest bands in the RR spectra which can be assigned to C=C stretching vibrations of the retinal moiety were found to be characteristic of each of the three chromophores and therefore were used for probing their time-dependent concentration. This type of RR detection was found to be more accurate than conventional optical detection since the fairly narrow vibrational bands can be identified and separated more precisely than the broad and strongly overlapping optical absorption bands. On this basis it was found as a novel and quite unexpected result that during the photocycle an intermediate equilibrium between L-550 and M-412 is established (L-550: M-412 $\approx$ 1:3 at pH 7.4) which finally decays with a time-constant of 5 ms. This is attributed to an oscillatory motion of a proton between the SB group and its counterion (Alshuth and Stockburger, 1985). Our results throw new light on the role of BR's photocycle for the proton pump mechanism.

REFERENCES

1. T. ALSHUTH, P. HILDEBRANDT and M. STOCKBURGER, in Spectroscopy of Biological Molecules, edited by C. Sandorfy and T. Theopanides, D. Reidel, Dordrecht, 329-346 (1984).
2. T. ALSHUTH and M. STOCKBURGER, Photochem. Photobiol., (submitted for publication) (1985).
3. M. STOCKBURGER, W. KLUSMANN, H. GATTERMANN, C. MASSIG and R. PETERS, Biochemistry 18, 4886 (1979).

CHAPTER 9

NON LINEAR SPECTROSCOPY OF CHLOROPHYLL DIMERS AND SOLID TRANS β-CAROTENE*

J. A. KONINGSTEIN
The Ottawa-Carleton Institute for Graduate Studies and Research in Chemistry,
Department of Chemistry,
Carleton University,
Ottawa,
Ontario K1S 5B6, Canada.

1. INTRODUCTION

During the last years we have carried out time resolved fluorescence and Raman studies of the biological interesting molecules Chlorophyll a Dimer and all trans β-carotene. The dimer molecules exist over a wide range of concentrations of dry chlorophyll a in dry hexane, carbon tetrachloride and benzene. On the other hand there are two molecules of trans β-carotene in the unit cell - space group $C^5{}_{2h}$ - and our interest is centred around the "pair" aspect of these systems. If compared to the energy level diagram of the individual molecules, that of the dimer of chlorophyll a and of trans β-carotene in the solid state should show a doubling of states. The splitting of these states for the dimer is thought to be associated with the

*Research supported by the Natural Sciences and Engineering Research Council of Canada.

existence of a pseudo exciton in the pair molecule where one molecule acts as an electron donor and the other as an acceptor.[1] They reverse roles at a time scale to fast to become detectable in N.M.R. but perhaps slow enough to reveal itself in optical spectroscopic studies. A search for the pseudo exciton splitting has been the subject of several investigations but they were not successful. The splitting of electronic states of solid trans β-carotene - as compared to the single electronic state of trans β-carotene itself - is called the Davydov splitting. Application of linear spectroscopic methods, that is one photon absorption, fluorescence and the common Raman effect do not permit the assignment of spectral features to either the pseudo-exciton or Davydov splitting. Presently we discuss the use of non-linear spectroscopic methods.

2. RESULTS AND DISCUSSION

2.1. Davydov Splitting and Spectra of Solid Trans β-Carotene.

Raman spectroscopy has proven to be a powerful tool as a method to study the vibrational spectrum of the trans β-carotene molecule in the solid state as well as in solution. The absorption spectrum in the visible is characterised by a band system due to transitions between $S_2(^1B_{1u})$ and $S_o(^1A_g)$ electronic surfaces - see Figure 1. The cross sections of these transitions is large ($\varepsilon \approx 10^5$) which means (i) a large resonance enhancement for Raman intensities of normal modes in the S_o ground state and (ii) the possibility of (S_2) carrying out excited state vibrational Raman spectroscopy with pulsed lasers. The

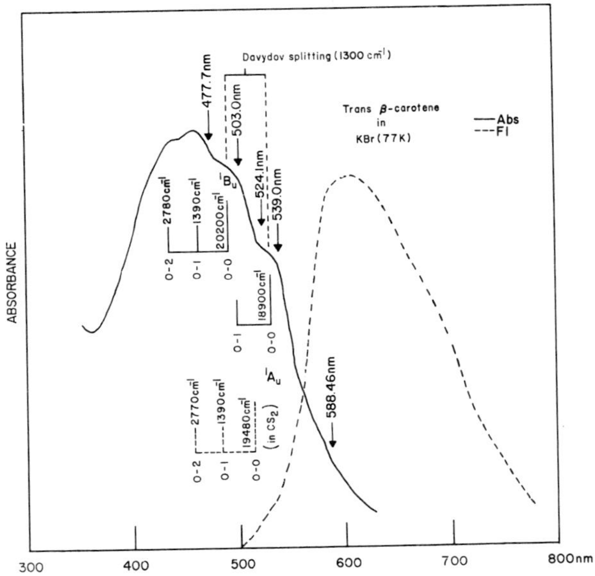

FIGURE 1 The absorption and fluorescence spectrum of trans β-carotene in KBr at 77 K.

lifetime of S_2 is short (<1 ps) and consequently the mono-beam pump/Raman probe technique can be applied.[2] This technique makes use of the fact that - using nanosecond pulsed lasers - the time evolution of the population of the short lived excited state closely follows the shape of the laser pulse. Other photons in the laser beam now can act as the Raman probe and pump and probe have a perfect overlap in space and time. For trans β-carotene in the solid state pumping followed by Raman probing has been carried out at wavelengths between 450 nm and 590 nm, see Figures 1 and 2. Our results reveal the following: pumping followed by Raman probing with wavelength 480 nm < λ_ℓ < 520 nm reveal that besides scattering from normal modes in the ground state also that of scattering of such modes in electronic excited states is present. For

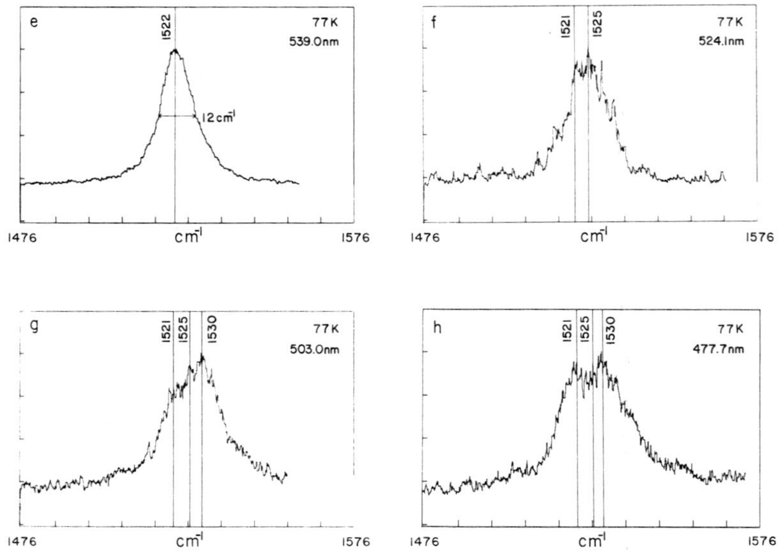

FIGURE 2 Raman scattering of the ν_1 mode (c=c stretch) at 1521 cm^{-1} in the ground state, at 1525 cm^{-1} in the ^1A_u and at 1531 cm^{-1} in the ^1B_u Davydov components.

instance - see Figure 2 - in the spectral interval with Raman shifts from 1476 cm^{-1} - 1576 cm^{-1} we observe bands at 1521 cm^{-1} and 1531 cm^{-1}. Time resolved Raman studies clearly show - see Figure 3 - that whereas the relative intensity of the 1531 cm^{-1} band increases in time, that of the 1521 cm^{-1} decreases. The latter band is due to scattering from a normal mode in the ground state and the former from the corresponding mode in an electronic excited state which must have a lifetime longer than the duration of the pulse. The time evolution of population of such a long lived excited state is retarded with respect to that of the laser pulse and hence vibrational Raman scattering from modes in this stage gain in intensity with respect to

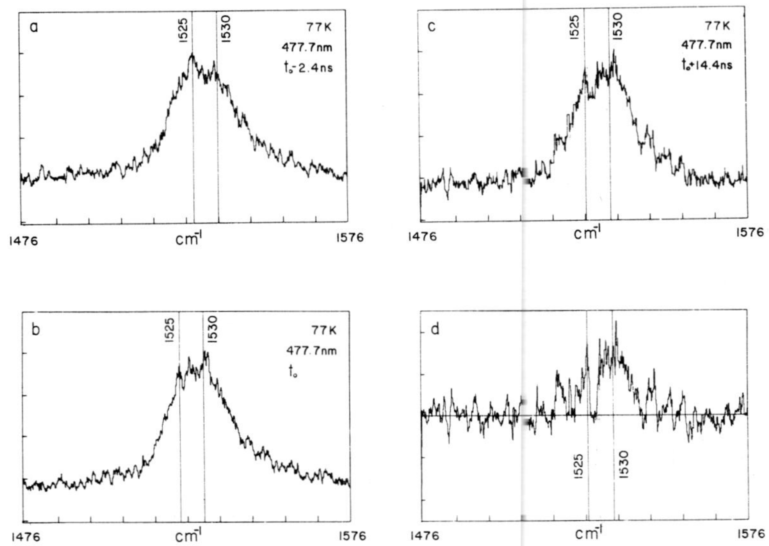

FIGURE 3 Time resolved Raman measurements of the ν_1 mode in the two Davydov components. The laser pulse peaks at t_o and d is the difference spectrum of a (before) and c (after the pulse).

that of scattering in the ground state if t increases. The ratio of population numbers of ground and electronic state is not at all related to the shape of the pulse. Pumping and probing with pulsed laser radiation shows a decrease in the contribution of vibrational scattering from this rather long lived excited state. However, contribution to the spectrum of scattering from another electronic excited state is also recorded. This contribution has a maximum value for <480 nm <520 nm and time resolved studies show that the lifetime of this excited state is <u>shorter</u> than those of the 4 ns pulse. The position of normal modes in the two different electronic

thought to be none fluorescent, however using the pulsed laser Raman set up we recently succeeded to record this fluorescence.

The absorption spectrum of monomeric Chlorophyll a in solution consists of a band in the red (the Q_y band) and the Soret band in the blue - see Figure 4. Another band (Q_x) is thought to be located between Soret and Q_y bands but its position cannot be located because this band is hidden in vibronic bands of Q_y. The absorption spectrum of the dimer - also shown in Figure 4 - is similar to that of the monomer, however the absorptions are broader and red shifted. This could be an indication that the

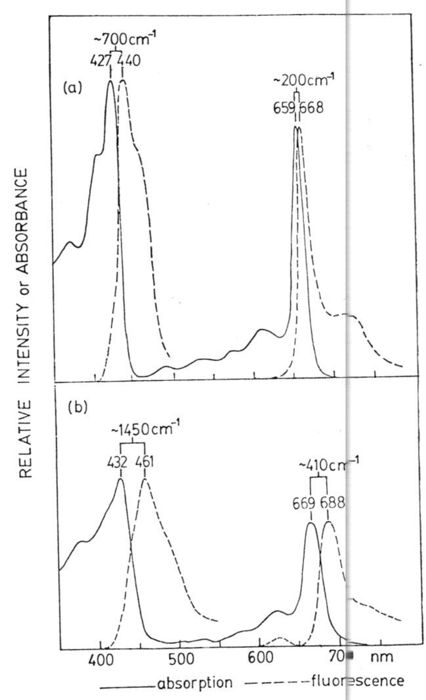

FIGURE 4 Absorption and fluorescence spectrum of Chlorophyll a in hexane. Top monomer, bottom dimer.

configuration for Chlorophyll a is that of a folded dimer. An analysis of the pulsed laser induced fluorescence spectrum[3] does reveal emission in the Q_x and Q_y spectral regions at ∿625 nm and 680 nm. The decay time of these emissions are in the nanosecond time region. Using two red photons, blue fluorescence is induced via a consecutive resonant two photon absorption process which involved the Q_y levels of the dimer. The red fluorescence of monomeric and dimeric Chlorophyll a induced by blue laser light can be quenched. The quenching process with blue photons relates to the fact that the nanosecond lived Q levels become populated - via vibrational relaxation - and the spontaneous red emission competes with stimulated absorption of blue photons from the Q levels.

It is also possible to excite a vibrational Raman spectrum of the dimer with these blue photons, which are in resonance with the dimeric Soret band - see Figure 5. An interesting feature is observed in the excitation profile of some of the normal modes. Rather than reaching a maximum, these excitation profiles reveal two sharp holes - Figure 6. These holes[4] are thought to be related to the quenching of red fluorescence mentioned above in the following way. It appears that the absorption of the blue photons from the dimeric Q level and/or from electronic states lying in the Soret band are responsible for the destruction of the dimer. The latter reveals itself experimentally in the form of a thin film which is deposited on the wall of the Raman cell. The destruction of the dimer is the result of a non linear absorption process of blue photons and this process is critically dependent on the position of electronic origins in the dimeric Soret band. Under conditions where population of

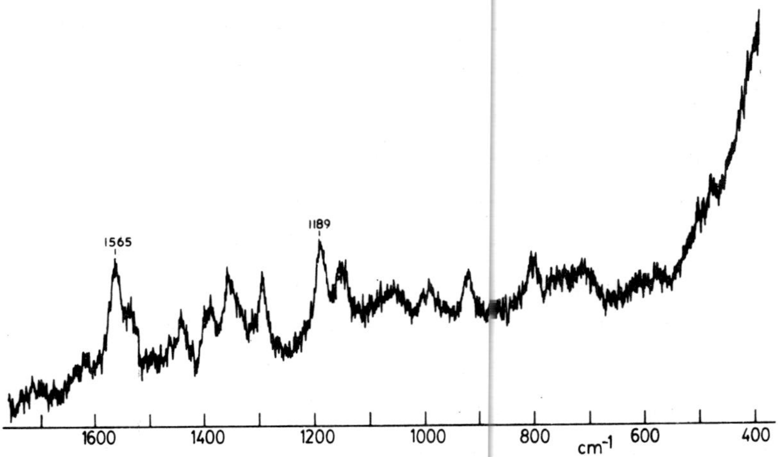

FIGURE 5 Pulsed laser induced Raman spectrum of the dimer in hexane. λ_ℓ = 433.6 nm.

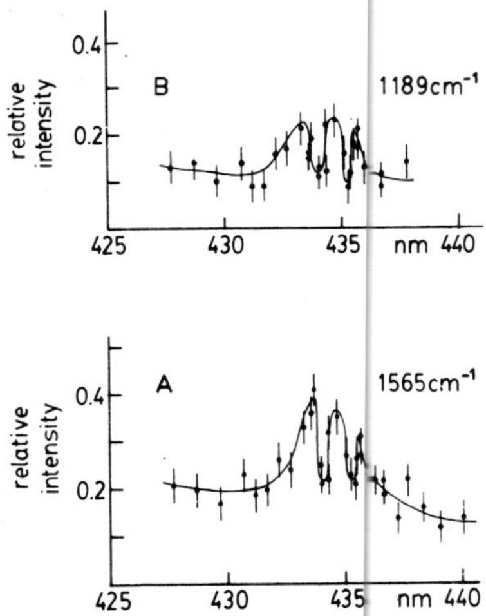

FIGURE 6 The excitation profiles of the intensity for two normal modes of the dimer.

excited states does not take place, the excitation profile of a Raman band of a normal mode should show a maximum at the wavelength of the laser in resonance with the position of the o-o transition. However, in the case of dimeric Chlorophyll *a* the dimer is destroyed if the laser is tuned towards the wavelength of the o-o transition(s), hence the holes in the excitation profiles. The composition of the Soret band of the monomer is rather complicated in that there are at least two electronic origins (B_x and B_y). These B levels are split in the dimer into at least an additional two components. The splitting of the latter is thought to be related to a pseudo exciton splitting. Because Q_x and Q_y for the dimer are not degenerate one expects this to be also the case for B_x and B_y. In that case it seems unlikely that the two holes which are apart by only 60 cm^{-1} are due to absorptions to B_x and B_y, rather they may well represent the pseudo exciton splitting of either B_x or B_y of the dimer.

3. CONCLUSIVE REMARKS

The experiments described above demonstrate that part of the energy level diagram of Chlorophyll *a* dimers and trans β-carotene can be constructed following the results of non linear studies which include time resolved excited level Raman and consecutive absorption spectroscopy. It is thought that the results of these studies are of relevance to an understanding of energy transfer between these molecules in solution or that of chlorophyll aggregates in β-carotene in the solid state. For instance pumping into the β-carotene bands in the blue and followed by the recording of the red fluorescence of the dimer could point to energy transfer between these molecules in solution. In a similar way the excitation profile of Raman band of these molecules could yield important information on the nature of the energy transfer process.

REFERENCES

1. J. J. KATZ, L. L. SHIPMAN, T. M. COTTON and T. R. JANSON, The Porphyrins, Vol. 5C, Ed. D. Dolphin, Academic Press, New York, pp. 401-58, (1978).
2. L. V. HALEY and J. A. KONINGSTEIN, Chem. Phys., 77, 1-9, (1979).
3. A. C. DE WILTON, L. V. HALEY and J. A. KONINGSTEIN, J. Phys. Chem., 88, 1077-83, (1984).
4. A. C. DE WILTON, L. V. HALEY and J. A. KONINGSTEIN, Chem. Phys. Letters, 97, 530-40, (1983).

Subject Index